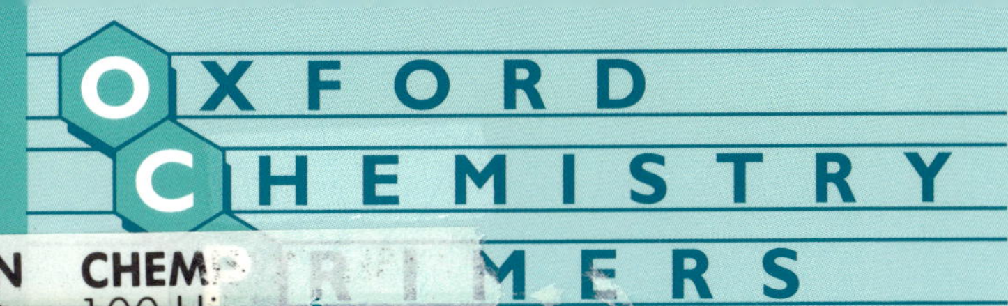

This series of short texts provides accessible accounts of a range of essential topics in chemistry and chemical engineering. Written with the needs of the student in mind, the Oxford Chemistry Primers offer just the right level of detail for undergraduate study, and will be invaluable as a source of material commonly presented in lecture courses yet not adequately covered in existing texts. All the basic principles and facts in a particular area are presented in a clear and straightforward style, to produce concise yet comprehensive accounts of topics covered in both core and specialist courses.

This book is a short and up-to-date text on oxidation and reduction, suitable for undergraduates studying organic chemistry. Redox reactions are a theme that runs through the core of organic chemistry. An appreciation of what constitutes an oxidation, or a reduction, is essential to gain an understanding of chemical reactivity.

A range of oxidation and reduction reactions is described, together with examples of these reactions being put to use in organic synthesis. The mechanisms by which oxidising and reducing agents operate are discussed, and this will lead to an appreciation of how to selectively oxidise, or reduce, one functional group in the presence of another.
In addition, asymmetric variants of key redox reactions are discussed as these constitute the cutting edge of chemical methodology.

Timothy Donohoe is Reader in Organic Chemistry at the University of Manchester.

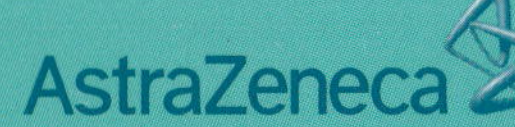

The low price of this book has been made possible by the generous sponsorship of AstraZeneca Limited

OXFORD UNIVERSITY PRESS

ISBN 0-19-855664

Oxidation and Reduction in Organic Synthesis

Timothy J. Donohoe

Reader in Organic Chemistry, University of Manchester

OXFORD
UNIVERSITY PRESS

Great Clarendon Street, Oxford OX2 6DP
Oxford University Press is a department of the University of Oxford.
It furthers the University's objective of excellence in research, scholarship,
and education by publishing worldwide in

Oxford New York

Athens Auckland Bangkok Bogotá Buenos Aires Calcutta
Cape Town Chennai Dar es Salaam Delhi Florence Hong Kong Istanbul
Karachi Kuala Lumpur Madrid Melbourne Mexico City Mumbai
Nairobi Paris São Paolo Singapore Taipei Tokyo Toronto Warsaw

with associated companies in Berlin Ibadan

Oxford is a registered trade mark of Oxford University Press
in the UK and in certain other countries

Published in the United States
by Oxford University Press Inc., New York

First published 2000

A catalogue record for this book is available from the British Library

Library of Congress Cataloging in Publication Data
(Data applied for)

ISBN 0 19 8556640 acv

Typeset by the author
Printed in Great Britain
on acid-free paper by Bath Press Ltd, Bath, Avon

Pam

Series Editor's Foreword

Changing the oxidation level is one of the most common processes encountered in organic synthesis: It is the mainstay of many elegant total syntheses and understanding how to achieve such changes, particularly of one functional group in the presence of others, is an essential part of a chemistry student's repertoire.

Oxford Chemistry primers have been designed to provide concise introductions relevant to all students of chemistry and contain only the essential material that would normally be covered in an 8–10 lecture course. This present Primer by Tim Donohoe presents the concepts and mechanisms of both oxidations and reductions in a very methodical and student friendly fashion. This primer will be of interest to apprentice and master chemist alike.

Professor Stephen G. Davies
The Dyson Perrins Laboratory,
University of Oxford

Preface

Pick a synthesis of an important organic compound from the literature: the chances are high that the route will contain some type of oxidation or reduction reaction. Indeed, the theme of oxidation and reduction runs through the very core of organic synthesis and it has been my aim to provide a guidebook to this area. In fact the spectrum of redox reactions is so great that this short text is designed only to describe briefly some of the more interesting and useful ones. I have highlighted the area of asymmetric synthesis using enantiopure catalysts as I think that some of these redox reactions represent the cutting edge of organic chemistry. I have also attempted to illustrate the types of selectivity (chemo- and stereoselectivity for example) that are possible under oxidising and reducing conditions, as this knowledge is essential when planning a synthesis.

In each reaction that is encountered I have endeavoured to provide a guide to a likely mechanism so that the book is more than just a list of oxidising and reducing agents. In some (in fact most) cases the mechanism of a reaction has not been proven beyond doubt and here I have tried to present a logical and reasonable interpretation of the results.

Finally, I would like to express my sincere gratitude to Drs J. P. Clayden, M. S. Loft and A. T. Russell who each proof-read the entire manuscript.

2000 T. J. D.

Contents

1. Introduction

f we look at the multitude of varied and interesting reactions that constitute organic chemistry, it is possible to classify a great many of them as either oxidation or reduction reactions. Oxidation and reduction is a theme that runs through the very core of organic chemistry and it is difficult indeed to think of a synthesis (either in the laboratory or in nature) which does not use such a reaction at one stage or another.

1.1 What are oxidations and what are reductions?

Before we examine details of the oxidation and reduction process it is instructive to consider some definitions which will serve us well later on.

Oxidation of an organic subtrate may be defined as:
The addition of oxygen,
OR
The removal of hydrogen,
OR
The removal of electrons , from that compound.

Similarly, **reduction** can be defined as:
The removal of oxygen,
OR
The addition of hydrogen,
OR
The addition of electrons, to an organic substrate.

General introduction to oxidation states and levels

Perhaps you will have come across the concept of oxidation states in inorganic chemistry, where metals can be placed into a variety of different categories depending upon how many electronegative ligands they are bonded to. For example, the chromium in chromium trioxide is easily categorised as being in the +6 oxidation state, while manganese in manganese dioxide is in the +4 oxidation state. This method of characterising metals is often quite useful when examining the reactivity of various complexes. For instance, the osmium atom in osmium tetroxide is formally in the +8 oxidation state (we assign this by giving the osmium atom +1 for every bond to a more electronegative element). This high state means that the complex is quite likely to be a good oxidising agent as osmium can commonly exist in the +3 or +4 oxidation state; this is the case and osmium tetroxide is a useful oxidant which we shall come across later. However, we should not imagine that because osmium is in the +8 oxidation state there are actually Os (8+)

ions in solution, waiting to be reduced: if there were, then osmium tetroxid would be the world's best Lewis acid and this is not the case.

The classification of organic compounds into oxidation states is mor difficult and inconsistencies abound; and although one can devise a system fc achieving this classification, it is not particularly useful to organic chemists For example, the simplest way to assign an oxidation state to a particula carbon atom is to count the ligands around it and give the carbon +1 for ever element more electronegative than carbon; 0 for every other carbon atom; – for every hydrogen atom. Using this system we shall consider hexane. C-1 c hexane has an oxidation state of –3 (–1 + –1 + –1 + 0) while the C-2 c hexane has an oxidation state of –2 (–1 + –1 + 0 + 0). Clearly there is problem here as most organic chemists would not think of these two carbo atoms as being in different oxidation states. Therefore, we need a mor flexible (and also less precise) method of classifying different carbon atom which will be useful to us as synthetic chemists. Importantly, it shoul classify various carbon atoms in groups which are of similar reactivity wit respect to redox reactions.

Oxidation levels in organic chemistry

Organic chemists find it convenient to place functional groups into fiv different categories (or levels) depending upon the level of oxidation containe within. These are rather broad definitions and there is some scope for orderin compounds within each level according to the rule described above: howeve this need not concern us.

It should also be noted that we are concerned with ordering of functiona groups themselves and that these are mostly (but not exclusively) centre around one carbon atom— alkenes and alkynes are obvious exceptions t this. Multi-functional organic compounds may therefore contain carbo atoms which are themselves in several different oxidation levels.

Level 0

The lowest oxidation level for carbon is 0 (which we shall call th hydrocarbon level). The carbon of methane and each of the carbons of butan are contained within this level, Fig. 1.1. Note that each carbon has *zer* bonds to elements more electronegative than itself.

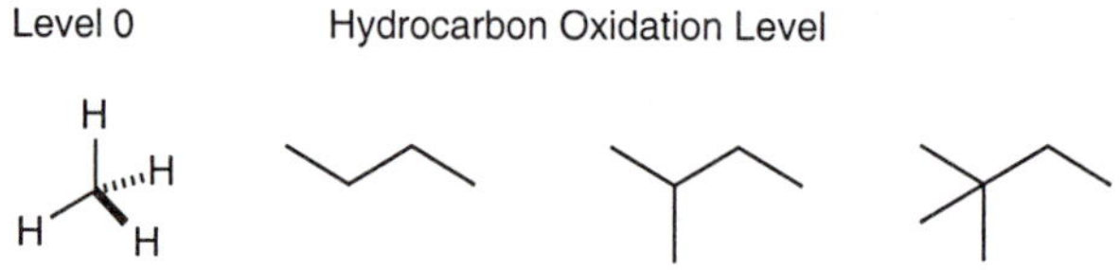

Level 0 Hydrocarbon Oxidation Level

H H H H

Fig. 1.1

Level 1

The next oxidation level is 1 (which we shall call the alcohol level). In thi category are carbon atoms which have *one* bond to an electronegative atom such as alcohols, thiols, alkyl halides, alkyl nitro compounds and amines. We shall also include the alkene functional group here, Fig. 1.2. It is also appropriate to include benzene as being in level 1.

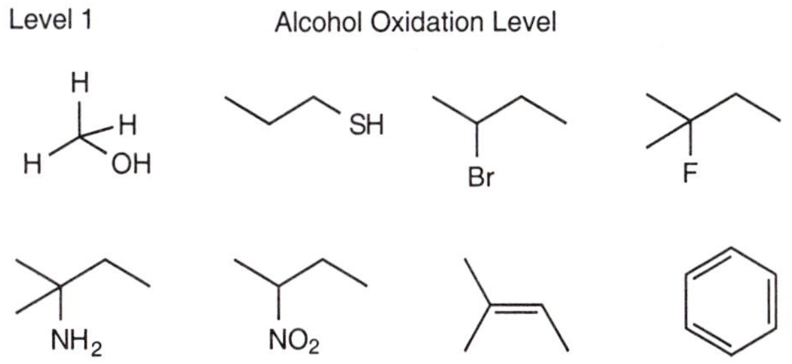

Level 1 Alcohol Oxidation Level

Fig. 1.2

The classification of alkenes in level 1 and alkynes in level 2 can seem surprising at first glance. However, consider that an alkene is readily transformed to an alcohol (level 1) under aqueous acidic conditions (of course the regiochemistry of this step depends on the alkene and the conditions used). Such addition of water constitutes neither an oxidation or a reduction (think about the elements of water, H_2 and O, cancelling each other out in a redox sense). The reverse process also illustrates this point as reaction of **1** with a dehydrating agent (say P_2O_5) gives an alkene. So, we can see that the alkene is equivalent to an alcohol and is rightly placed in level 1. Remember that in this case the oxidation level is equally shared between the two carbon atoms.

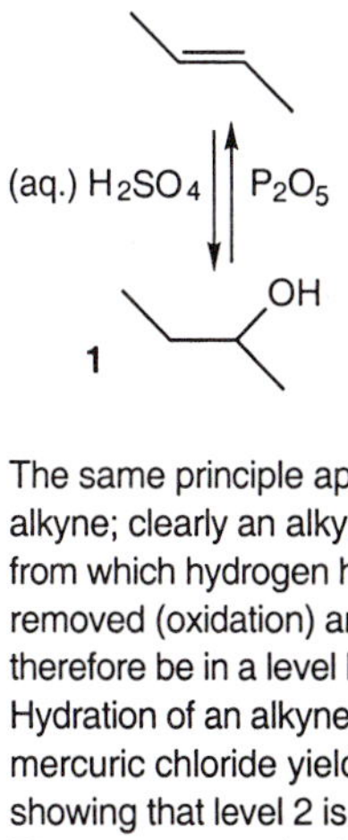

(aq.) H_2SO_4 P_2O_5

1

Level 2

The third oxidation level, 2 (which we shall call the ketone oxidation level), contains carbon with *two* bonds to electronegative atoms, such as acetals, thioacetals, and geminal dihalides. As an obvious extension of this, we shall also include aldehydes, ketones, and imines. This grouping also contains alkenes that are substituted with one electronegative atom, such as enol ethers; the alkyne functional group is also in this level, Fig. 1.3.

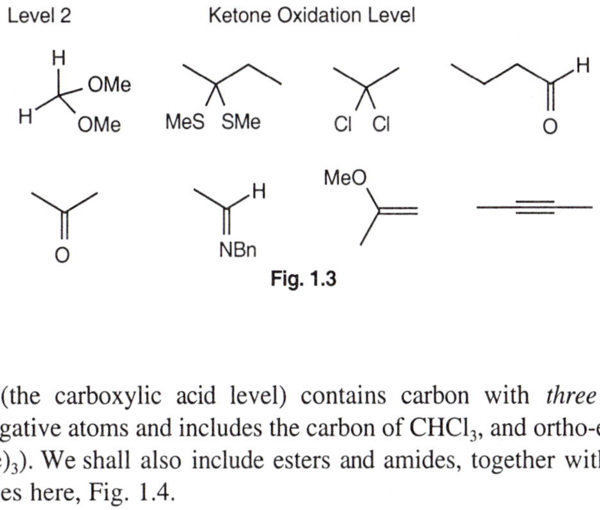

Level 2 Ketone Oxidation Level

Fig. 1.3

The same principle applies for an alkyne; clearly an alkyne is an alkene from which hydrogen has been removed (oxidation) and should therefore be in a level higher than1. Hydration of an alkyne with aqueous mercuric chloride yields a ketone showing that level 2 is the correct one. The reverse reaction is not particularly easy to perform but it can be achieved in a few steps without either an oxidation or a reduction.

(aq.) H_2SO_4 $HgCl_2$

Level 3

Level 3 (the carboxylic acid level) contains carbon with *three* bonds to electronegative atoms and includes the carbon of $CHCl_3$, and ortho-esters (e.g. $CH(OMe)_3$). We shall also include esters and amides, together with imidates and nitriles here, Fig. 1.4.

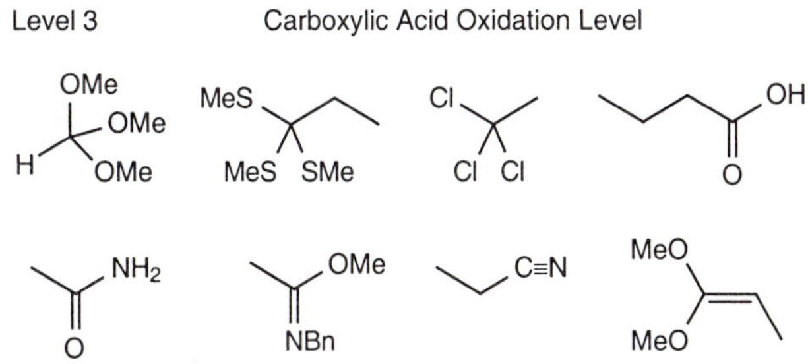

Level 3 Carboxylic Acid Oxidation Level

Fig. 1.4

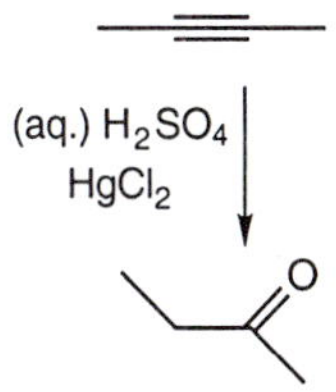

Level 4

Finally, the most oxidised level of carbon is 4 (carbon dioxide level). Obviously, carbon bonded to *four* electronegative ligands, such as in carbon tetrachloride, is included. As long as we have four bonds to electronegative atoms, we can have a mix of singly and doubly bonded groups such as ureas (e.g. NH_2CONH_2), carbamates (e.g. NH_2CO_2R), and carbonates (e.g. $(MeO)_2CO$). We would also want to add cyanamide (NH_2CN) to this category; finally, carbon dioxide, disulfide and oxysulfide are also included, Fig. 1.5.

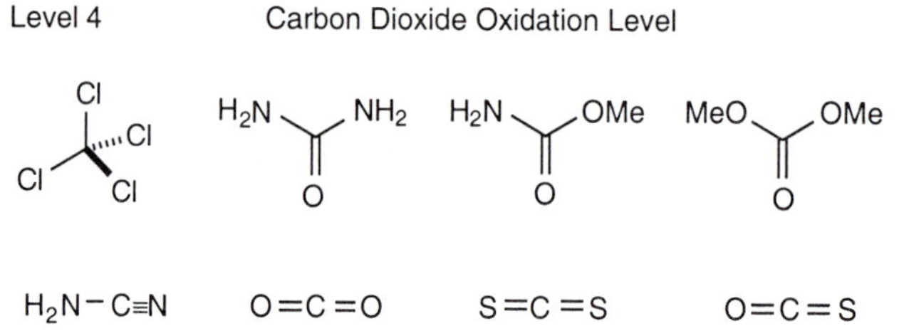

Level 4 Carbon Dioxide Oxidation Level

CCl_4 $H_2NC(O)NH_2$ $H_2NC(O)OMe$ $MeOC(O)OMe$

$H_2N-C{\equiv}N$ $O{=}C{=}O$ $S{=}C{=}S$ $O{=}C{=}S$

Fig. 1.5

The important point to remember is that one can move *within* a particular level by performing chemistry that is classified as neither an oxidation nor a reduction. However, you will find that it is not possible to move *between* levels without performing a reduction (to move down levels) or an oxidation (to move up levels).

It is also essential that you understand that these levels are rather loose definitions meant as a teaching aid rather than as being a comprehensive and rigorous classification of all known organic functional groups.

Some examples follow to help you understand this cataloguing.

Consider *n*-butane, **2**, Fig. 1.6. If one were to perform a free radical chlorination then it is possible to isolate 2-chlorobutane. One carbon of the butane skeleton has now been oxidised to the alcohol level (and therefore we should consider chlorine as an oxidising agent). If we take 2-chlorobutane and react it with aqueous acid then the result is formation of 2-butanol *via* a nucleophilic substitution. We are simply moving within a level during this reaction as both starting material and product are in level 1; therefore this is not a redox reaction.

2-Butanol can, however, be oxidised further (change to level 2) by reaction with chromium trioxide; the product from this reaction is butanone. Note, chromium(VI) is reduced to chromium(IV) in the process. Reaction of the ketone with methanethiol and $HgCl_2$ gives the corresponding thioacetal in a reaction that maintains the oxidation level at the functionalised carbon (starting material and product are both in level 2).

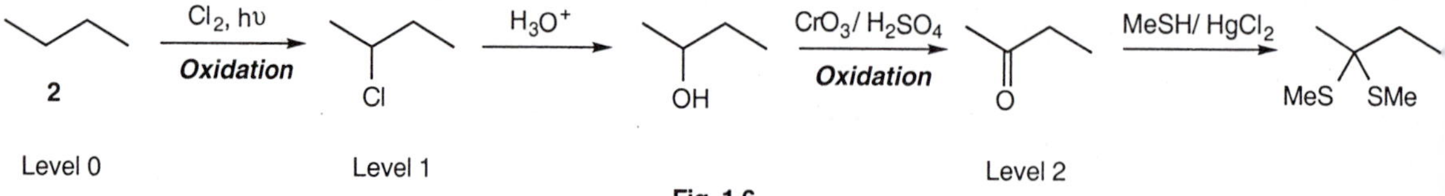

2 —(Cl_2, hυ; Oxidation)→ Cl —(H_3O^+)→ OH —(CrO_3/ H_2SO_4; Oxidation)→ O —(MeSH/ $HgCl_2$)→ MeS SMe

Level 0 Level 1 Level 2

Fig. 1.6

Study this example and you should be able to see another guiding principle concerning redox reactions, namely that *for every oxidation there is a corresponding reduction in the system* (in step one chlorine gas is reduced to chloride and in step three chromium is reduced) and *vice versa*.

Imagine another set of reactions, starting with valeronitrile, Fig. 1.7. Acid catalysed hydrolysis of the nitrile under standard conditions is merely a sideways move within level 3; the final step is a reduction, using $LiAlH_4$. In fact, reduction of a carboxylic acid to an alcohol with $LiAlH_4$ shows that it is possible to jump more than one level at a time as the starting material is in level 3 and the product is in the 1 oxidation level.

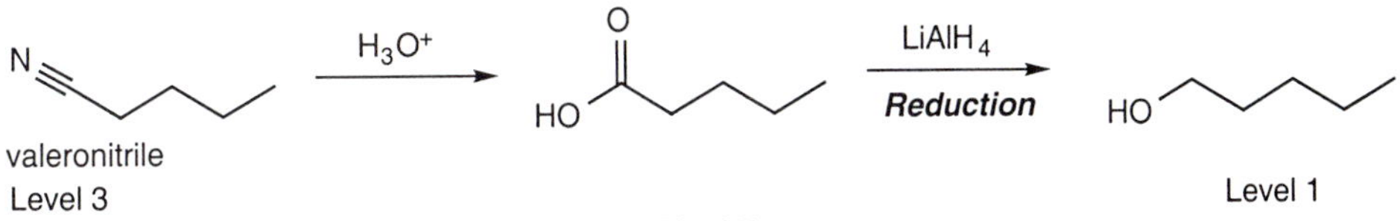

valeronitrile
Level 3
H_3O^+
HO
LiAlH$_4$
Reduction
HO
Level 1

Fig. 1.7

The concluding example in this chapter is designed to illustrate how we can classify functional groups that span more than one carbon atom, Fig. 1.8. Elimination of bromide **3** with a base is neither an oxidation nor reduction and can be seen as a sideways move within oxidation level 1. However, reaction of the product from the elimination reaction, an alkene, with bromine leads to an addition reaction that we shall discuss in more detail later. Suffice it to say at this point that the product is a vicinal dibromide **4** which is not in the same oxidation level as the alkene from which it originated— we have performed an oxidation on the alkene (Fig. 1.8). In fact, we can classify **4** as containing one functional group (a vicinal dibromide) in the 2 oxidation level. Now, consider elimination of two moles of HBr from the dibromide (use a strong base) to form alkyne **5**. This is just a sideways move within level 2.

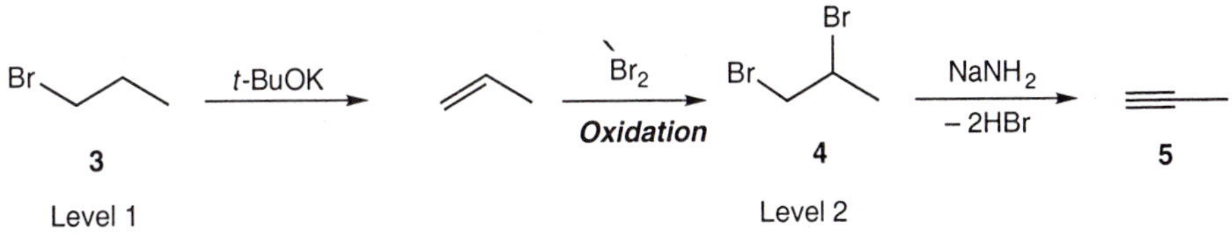

Br
3
Level 1
t-BuOK
Br_2
Oxidation
Br
Br
4
Level 2
$NaNH_2$
– 2HBr
5

Fig. 1.8

We need to be careful in trying to take this idea too far: for example, it is not helpful to think of 1,3-dibromopropane in the same way as we have classified 1,2-dibromopropane. In 1,3-dibromopropane, the two functional groups behave, to all intents and purposes, independently and the molecule should be classified as having two functionalised carbons, each in level 1. On the other hand, we have seen that the two functional groups in **4** can react in unison to form an alkyne: it is this ability that makes their joint classification in level 2 possible.

2. Oxidation of heteroatoms attached to carbon

It seems sensible to include the redox chemistry of some common heteroatoms in this book as the functional groups that are so formed are extremely useful in organic synthesis.

2.1 Oxidation of nitrogen

In order to introduce the oxidation reaction in earnest we should consider the reaction of an amine with an oxidant: what could be simpler? Unfortunately, the oxidation of amines can be a complicated reaction when primary and secondary amines are involved. However, we shall not consider these reactions in detail as the products (which are imines, nitroso compounds, nitro compounds, and oximes) are difficult to produce selectively. Instead, let us concentrate on the oxidation of tertiary amines to give the corresponding *N*-oxide. You should have no trouble in assigning this transformation as an oxidation because the product from the oxidation of **1** clearly contains more oxygen than it did before the reaction, Fig. 2.1. The mechanism of this reaction is illustrated by the oxidation of *N*-methylmorpholine with hydrogen peroxide and involves nucleophilic attack of the amine lone pair of electrons onto the peroxide, breaking the relatively weak oxygen–oxygen bond. Loss of a proton then furnishes the amine-*N*-oxide plus water, Fig. 2.1. Remember the structure of *N*-methylmorpholine-*N*-oxide **2** as it is an important oxidant in its own right (see Chapter 3) We can also see that the oxidant (hydrogen peroxide is a very simple and efficient oxidant for this purpose) is reduced to water, thus completing the redox cycle.

Remember that amine-*N*-oxides are zwitterionic structures in which there is effectively no π-bonding between the oxygen and the nitrogen (this is because nitrogen has no empty orbitals available for such π-bonding).

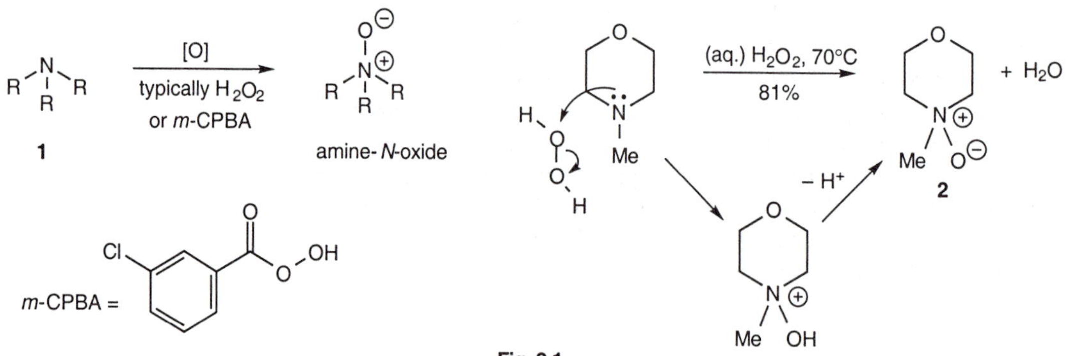

[O]
typically H_2O_2
or *m*-CPBA
1
amine-*N*-oxide
(aq.) H_2O_2, 70°C
81%
+ H_2O
$-H^+$
Me
Me OH
2
m-CPBA =

Fig. 2.1

In comparison with aliphatic amines, the oxidation of heterocyclic aromatic amines is less easy to accomplish with hydrogen peroxide—probably as a consequence of the trigonal nitrogen's reduced nucleophilicity. The best way to accomplish the oxidation of pyridines (for example) is to utilise *meta*–chloroperbenzoic acid (Fig. 2.2). As might be expected, electron-rich pyridines are more easily oxidised than electron–deficient ones. One must also be cautious about the oxidation of 2- and 2,6-substituted pyridines as the extra steric hindrance introduced by the *ortho* groups retards oxidation. This oxidation is quite a useful reaction as pyridine-*N*-oxides have a rich chemistry of their own— not least of which is the ability to allow electrophilic aromatic substitution on the heterocycle.

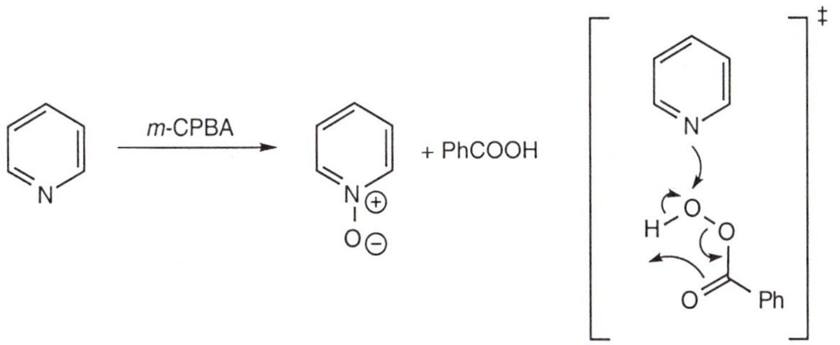

Recently a new oxidant has appeared which catalyses the (otherwise) slow reaction of a pyridine with hydrogen peroxide. Methyl trioxorhenium (MTO) very successfully promotes the oxidation of a range of both electron-rich and electron-deficient pyridines. The active oxidant may be the peroxy complex shown (formed from MTO and H_2O_2) which adds oxygen to the pyridine nitrogen.

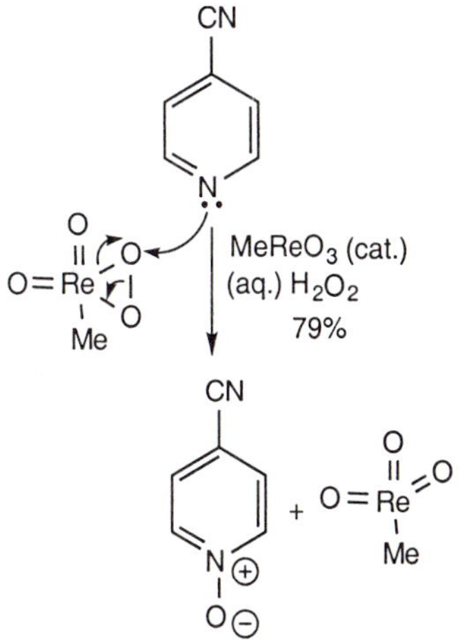

m-CPBA
+ PhCOOH
N
N⊕
O⊖
H
O
O
O
Ph
‡

Fig. 2.2

CN
N
$MeReO_3$ (cat.)
(aq.) H_2O_2
79%
O=Re
Me
CN
N⊕
O⊖
+ O=Re
Me

Oxidation of amines using a chiral oxidant

The combination of an oxidising agent with a chiral (non-racemic) ligand and a transition metal catalyst results in the formation of a chiral oxidant. And, as might be expected, chiral oxidants are capable of controlling the stereochemistry of an oxidation reaction (this will be treated in more detail in instances where a new stereogenic centre is formed during an oxidation reaction). However, another possibility for stereochemical control arises, even in oxidations that do not result in the formation of a new stereogenic centre. Consider the case of a substrate for oxidation that is itself chiral (and racemic). Here, one enantiomer of the substrate may well react with a chiral oxidant at a different rate than the other. In an ideal world, we would be able to stop the reaction at 50% conversion and recover a single, unreacted, enantiomer of substrate together with 50% of product which would be the oxidised (and therefore chemically different) form of the other enantiomer. This process is called kinetic resolution and is exemplified in Fig. 2.3. In this case, one enantiomer of the racemic amine **3** is unreactive while the other is oxidised to its *N*-oxide. Take note of the reagents used to construct the chiral oxidant: *t*-butyl hydroperoxide (source of oxygen), diisopropyl tartrate (chiral ligand) and titanium tetraisopropoxide (metal catalyst). A powerful chiral oxidant arises from the formation of an ordered complex between the metal, ligand and peroxide and it is this complex that is responsible for the kinetic resolution.

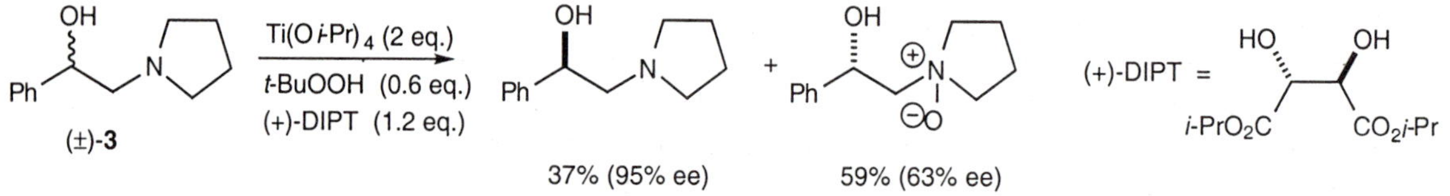

(±)-**3**

Ti(O*i*-Pr)$_4$ (2 eq.)
t-BuOOH (0.6 eq.)
(+)-DIPT (1.2 eq.)

37% (95% ee)

59% (63% ee)

(+)-DIPT =

Fig. 2.3

2.2 Oxidation of phosphorus

A pattern of reactivity similar to that of nitrogen can be observed in the oxidation of phosphorus–containing compounds. However, as phosphorus is lower down the periodic table than nitrogen it is generally more easily oxidised, as it is more nucleophilic. Consider the oxidation of phosphines to their corresponding oxides (Fig. 2.4). This reaction is affected by the steric hindrance around the phosphorus atom and so triphenylphosphine is stable in the atmosphere and requires a reagent such as hydrogen peroxide or manganese dioxide to produce triphenylphosphine oxide. On the other hand, trimethylphosphine is many times more reactive and will oxidise with just air itself. As might be expected, tributylphosphine displays reactivity somewhere between the two extremes.

The P–O bond length in triphenylphosphine oxide is rather short (1.43Å) suggesting the presence of substantial pπ–dπ back–bonding between the oxygen and empty d orbitals on the phosphorus. Thus, the double-bonded structure shown is a valid representation of the structure of phosphine oxides as well as the zwitterionic canonical (reality is somewhere between the two).

$R_3P{=}O \longleftrightarrow R_3\overset{\oplus}{P}{-}\overset{\ominus}{O}$

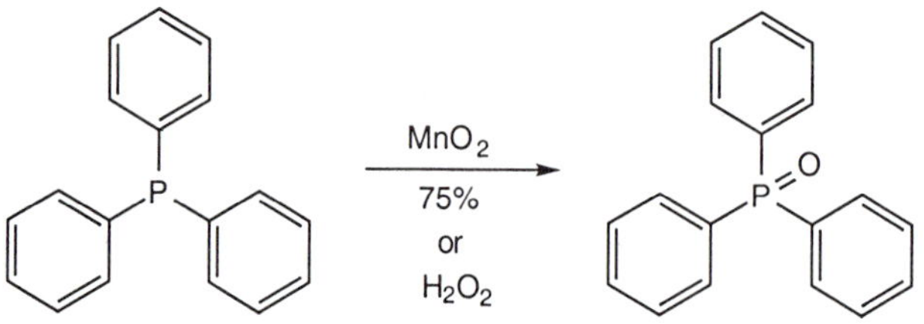

MnO_2
75%
or
H_2O_2

Fig. 2.4

Phosphine oxides are useful intermediates in alkene synthesis as α–alkoxy phosphine oxides undergo a *syn* elimination when treated with base. This sequence of events was utilised very successfully in the synthesis of *trans*-cycloctene from the epoxide (Fig. 2.5).

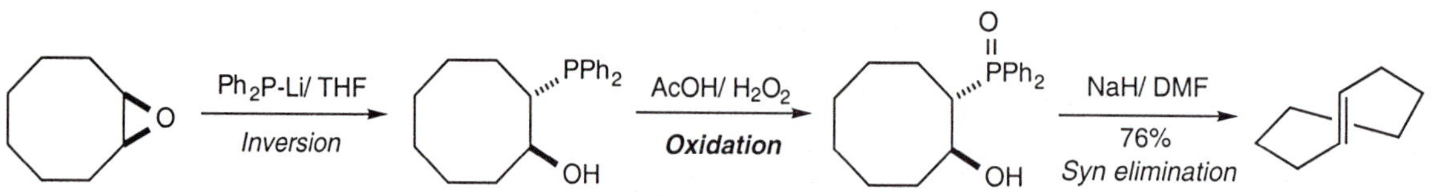

Ph_2P-Li/ THF
Inversion

AcOH/ H_2O_2
Oxidation

NaH/ DMF
76%
Syn elimination

trans-cyclooctene

Fig. 2.5

Although not strictly under the purview of this book (no carbon–phosphorus bonds are involved!) there is another oxidation of a phosphorus–containing group that is particularly useful to organic chemists. The phosphoramidate intermediates shown are used in the solid phase synthesis of oligonucleotides. Two sugar units are conveniently joined *via* reaction of a phosporamidate linkage with a free hydroxyl group (catalysed by a weak acid),

Fig. 2.6. After coupling, the phosphite triester product is oxidised with aqueous iodine to yield a phosphate triester which is now ready for coupling with another sugar unit (after deprotection of the P^1 protecting group). A mechanism for this oxidation is shown below. At the end of the synthesis, deprotection of all the phosphate ester protecting groups (P^2) takes place in one step to reveal the corresponding phosphate linkers between the sugars.

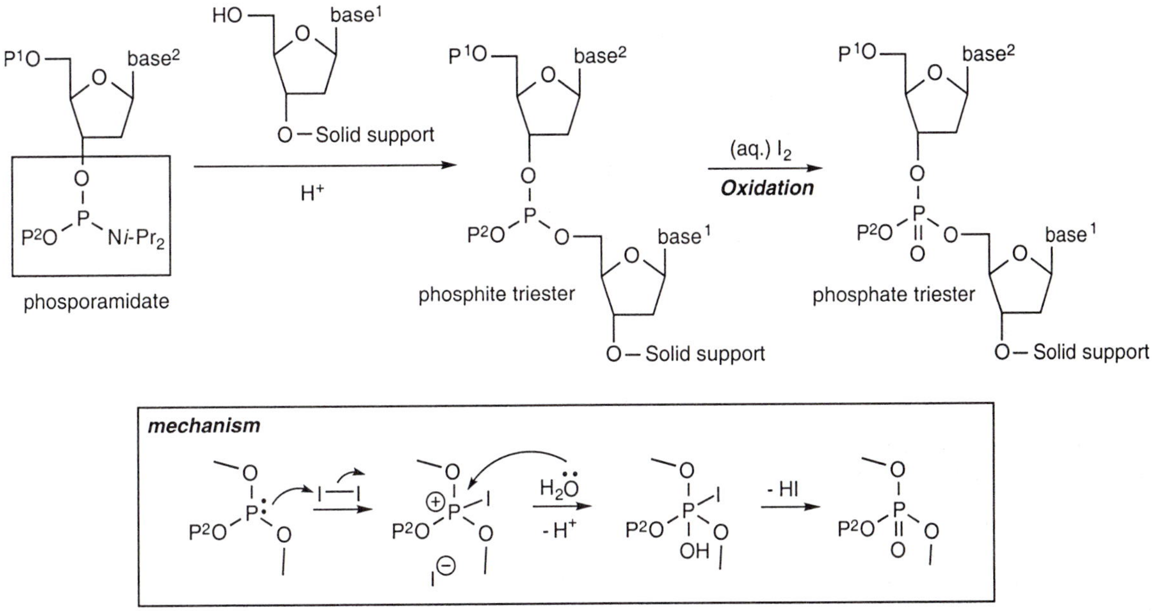

Fig. 2.6

2.3 Oxidation of sulfur and selenium

It is the oxidation chemistry of sulfur and selenium that deserves the most attention in this chapter. Both of these elements appear in functional groups with varying levels of oxygenation; as might be expected, they are both rather readily oxidised, with selenium being more easily oxidised than sulfur. We shall begin with the oxidation of thiols and selenols.

Formation of disulfides and diselenides

Selenols (RSeH) are not normally stable in air as they are readily oxidised to the corresponding diselenide (RSeSeR) by molecular oxygen. Thiols are somewhat more stable, although they too can be oxidised (to disulfides, RSSR) by oxygen if a base is present.

The oxidation of sulfides by air in the presence of base may proceed *via* the thiolate anion. This then undergoes a single electron transfer reaction with triplet oxygen to give a sulfur radical which can then dimerise.

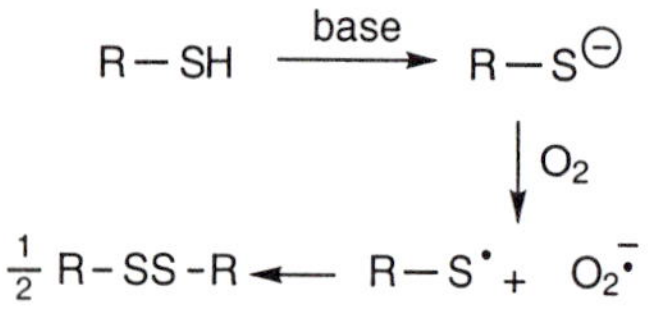

$$R{-}SH \xrightarrow{\text{base}} R{-}S^{\ominus} \xrightarrow{O_2} R{-}S^{\bullet} + O_2^{\bullet -} \longrightarrow \tfrac{1}{2}\, R{-}SS{-}R$$

Other oxidising agents such as hydrogen peroxide, manganese dioxide, and chromium trioxide may be used for the preparation of symmetrical disufides from thiols, although care must be taken to avoid overoxidation of the disulfide product. A nice way of obtaining unsymmetrical disulfides is shown in Fig. 2.7. Thus, reaction of ethane thiol with diethylazodicarboxylate (DEAD) furnishes a 1:1 adduct **4** which can then be treated with another (different if need be) thiol. Nucleophilic attack of this thiol onto the sulfur atom of the adduct ensues and the product disulfide is produced in excellent yield. Note that in the overall reaction the two thiols have been oxidised (one hydrogen atom has been removed from each sulfur) and that DEAD has been reduced.

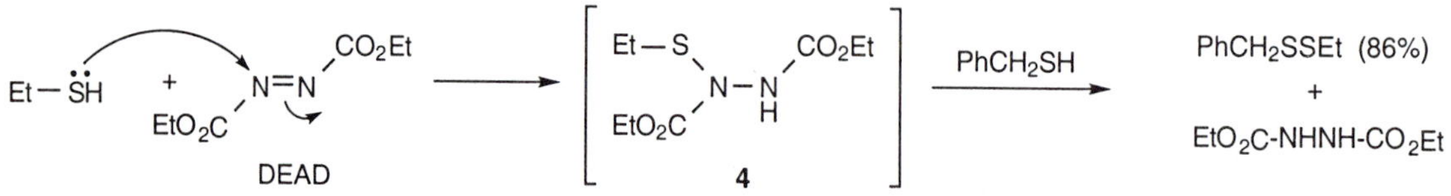

Et–SH + EtO$_2$C–N=N–CO$_2$Et (DEAD) → [Et–S–N(CO$_2$Et)–NH–CO$_2$Et] **4** —PhCH$_2$SH→ PhCH$_2$SSEt (86%) + EtO$_2$C-NHNH-CO$_2$Et

Fig. 2.7

Formation of sulfoxides and selenoxides

Sulfoxides and selenoxides are tetrahedral in shape, with two carbon groups, an oxygen and a lone pair of electrons at the four vertices. When the two carbon groups are non-equivalent ($R \neq R'$) then the sulfoxide is chiral (and configurationally stable).

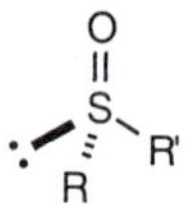

Oxidation of sulfides and selenides can give good yields of their corresponding oxides (namely, sulfoxides and selenoxides). Just like the phosphine oxides mentioned earlier, there is a reasonable amount of $p\pi$–$d\pi$ back–bonding in these species. There are many different oxidising agents which have been used for the oxidation reaction including hydrogen peroxide, *m*-CPBA, sodium periodate (NaIO$_4$), and ozone (O$_3$), Fig. 2.8. Although oxidation of sulfides to sulfoxides can be accomplished in the presence of other oxidisable groups such as pyridines and alkenes (with H_2O_2), care must always be taken to ensure that excess oxidant is not present, otherwise overoxidation of the sulfoxide to the sulfone can occur. The mechanism of oxidation with *m*-CPBA involves nucleophilic attack of the sulfide onto the peracid: if this is to be believed then there should be a correlation between the nucleophilicity of a sulfide and its ease of oxidation. This is indeed the case and, for example, dialkyl sulfides are oxidised more rapidly than their more hindered (and electronically deactivated) diaryl counterparts.

Sulfur acting as a nucleophile with *m*-CPBA

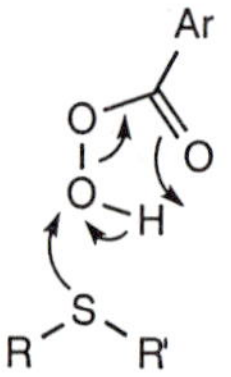

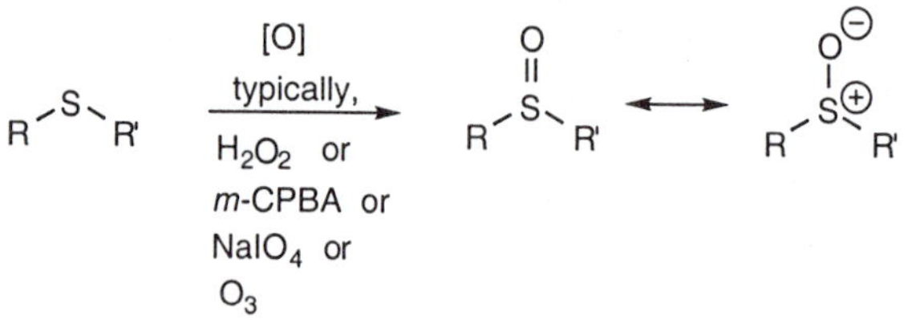

R–S–R' —[O] typically, H_2O_2 or *m*-CPBA or NaIO$_4$ or O$_3$→ R–S(=O)–R' ⟷ R–S$^{\oplus}$(–O$^{\ominus}$)–R'

Fig. 2.8

Interestingly, in most cases, it is not possible to isolate alkyl-selenoxides as they undergo a spontaneous *syn*-elimination process to form an alkene; this reaction is a particularly convenient way of introducing unsaturation adjacent to a carbonyl group, Fig. 2.9.

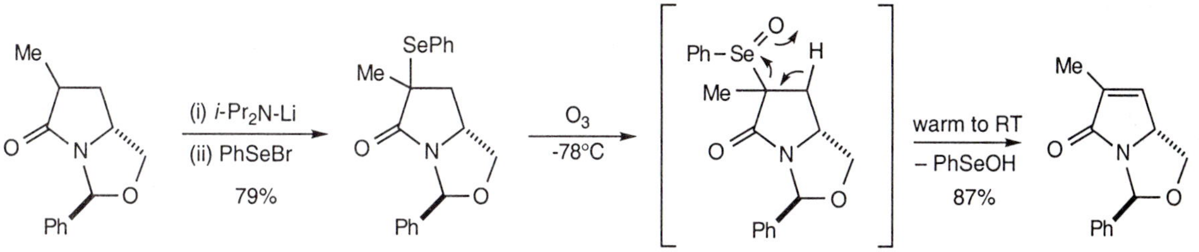

Fig. 2.9

Oxidation of sulfides using a chiral oxidant

Oxidation of unsymmetrical sulfides to the corresponding sulfoxides involves the formation of a new stereogenic centre (at sulfur). All of the oxidising reagents listed in Fig. 2.8 are achiral and therefore give a racemic mixture of products. However, oxidation using a chiral oxidant can give high selectivity for formation of a single enantiomer of product, Fig. 2.10. In the case shown it is the (+)-(*R,R*)-diethyl tartrate (DET)/ titanium tetraisopropoxide/ *t*-butyl hydroperoxide oxidising system that again proves to be very efficient at producing enantiomerically pure products.

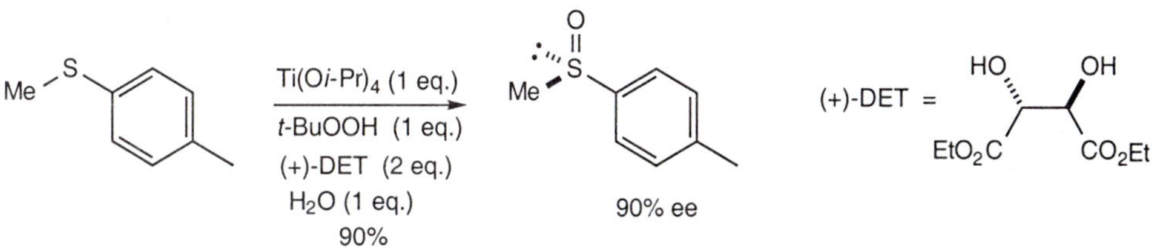

Fig. 2.10

Production of enantiomerically pure selenoxides is more difficult than that of the corresponding sulfoxides: this is because selenoxides are thought to racemise when exposed to water. The racemisation reaction takes place *via* the hydrate shown below. Once this is formed, any chirality that was present in the original selenoxide is lost.

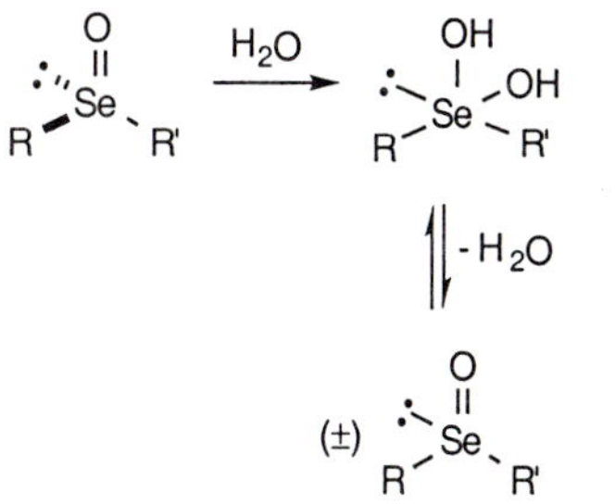

Formation of sulfones and selenones

Sulfur and selenium are each a part of another, even more oxidised functional group that is important in organic synthesis: these are named sulfones (RSO_2R) and selenones ($RSeO_2R$) respectively.

It is fortunate that the oxidation of sulfides to sulfoxides can be controlled so as to stop at the sulfoxide stage. This is because the oxidation of sulfoxides to sulfones is generally slower than the oxidation of sulfides to sulfoxides (by a factor of 10^2–10^3). This reactivity difference is to be expected with oxidants such as *m*-CPBA which are electrophilic— sulfoxides are bound to be less nucleophilic than sulfides as they are more hindered and also have an electron–withdrawing oxygen attached to the sulfur atom. So, if we want to complete the oxidation of sulfides to sulfones then we need to add excess oxidant, Fig. 2.11.

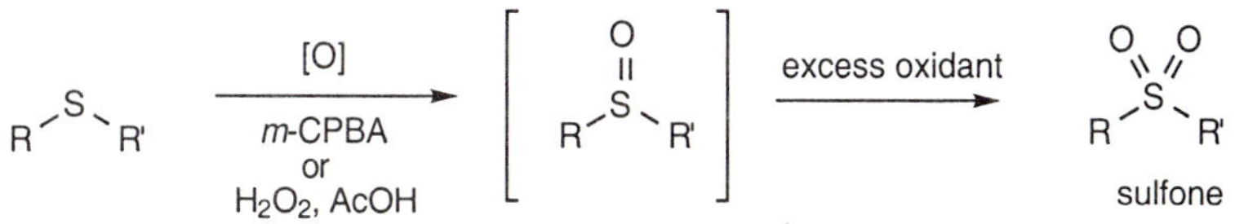

Fig. 2.11

Despite their reduced reactivity towards *m*-CPBA, sulfoxides can still be oxidised in the presence of other sensitive functional groups, such as alkenes and carbonyls, Fig. 2.12.

Acid–sensitive functionality can be preserved during sulfide oxidation by using sodium periodate.

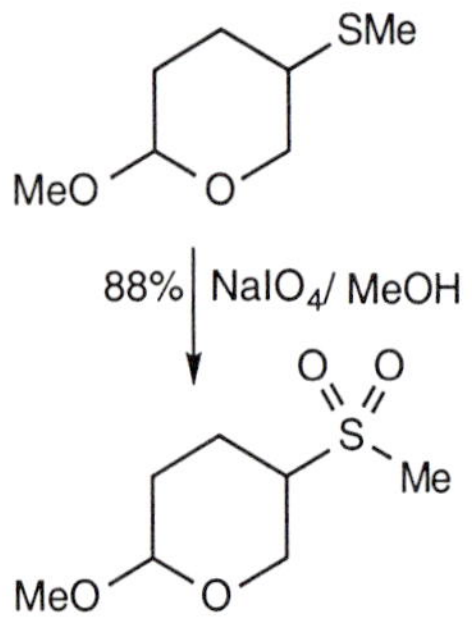

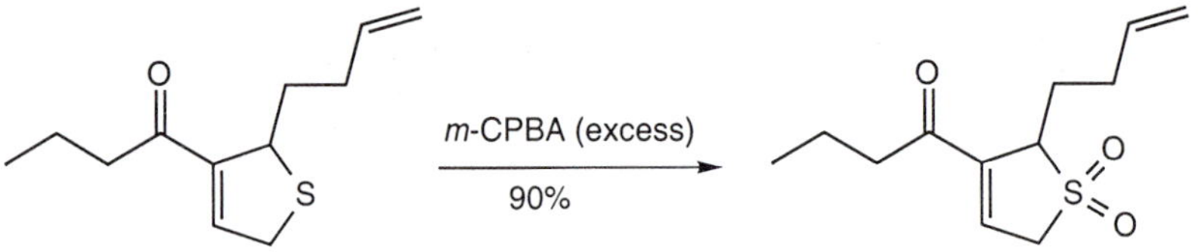

88% $NaIO_4$/ MeOH

m-CPBA (excess)

90%

Fig. 2.12

The relative rates of oxidation of sulfides and sulfoxides are reversed if one uses potassium permanganate as an oxidant. Presumably, this (negatively charged, MnO_4^-) oxidant is acting as a nucleophile, preferring to react first with an electrophilic sulfoxide rather than a sulfide. Using this reagent it is possible to oxidise sulfoxides to sulfones in the presence of a sulfide, Fig. 2.13. Compare this with the result obtained with *m*-CPBA at 0°C. Note that both the *bis*-sulfoxide and the mono-sulfone can be oxidised to the *bis*-sulfone by reaction with excess *m*-CPBA at elevated temperatures.

As might be expected, the *bis*-sulfoxide **5** was formed as a mixture of diasteroisomers.

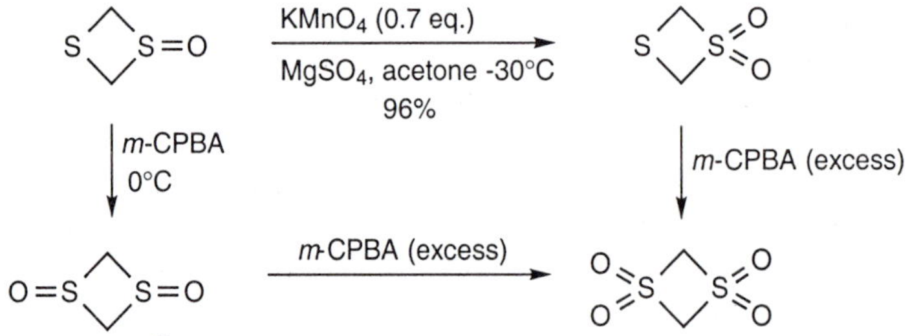

$KMnO_4$ (0.7 eq.)

$MgSO_4$, acetone -30°C

96%

m-CPBA

0°C

m-CPBA (excess)

m-CPBA (excess)

5

Fig. 2.13

The oxidation of selenides to selenones (*via* a selenoxide) is rather a difficult one to achieve and it appears to be the general instability of selenoxides (as noted above) that is mostly to blame. Perhaps the easiest group of selenones to prepare and study are those substituted with two aryl groups (β-elimination of any selenoxide intermediate is therefore precluded). Figure 2.14 shows some examples of selenone formation by oxidation of selenoxides. The direct conversion of aryl-alkyl selenides to selenones can be achieved with pertrifluoroacetic acid (CF_3CO_3H).

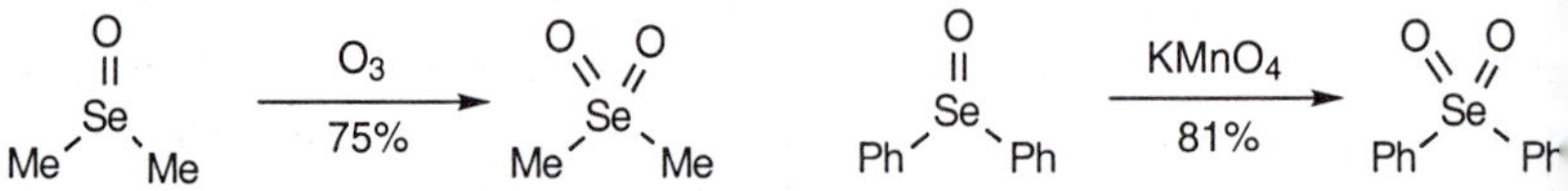

O_3

75%

$KMnO_4$

81%

Fig. 2.14

3. Oxidation of carbon–carbon double and triple bonds

'he oxidation of carbon–carbon π-bonds can yield several useful functional roups. Many oxidising agents have been developed for this purpose and we hall consider the most useful ones below. In fact, one of the main dvantages of oxidising alkenes is that the product distribution can be ontrolled, and a particular functional group formed, simply by the choice of xidant.

.1 Oxidation of alkenes to form epoxides

;poxides are the cyclic ethers that may be formed from the reaction of an lkene with an oxidising agent, Fig. 3.1. Remember that epoxides are lectrophiles and that (when protonated) they react with nucleophiles such as vater. The product from this reaction is a 1,2-diol (or vicinal diol). We can onsider the 1,2–diol as a single functional group, see page 5, which is in xidation level 2. As the epoxide furnishes the 1,2–diol simply by the ddition of water, we can consider this functional group as also being in level . Therefore, the transformation of an alkene (oxidation level 1) to an epoxide ; rightly considered as an oxidation reaction and the acid catalysed reaction of n epoxide with water is a sideways move within oxidation level 2.

The reaction of epoxides with a nucleophile takes place with inversion. The nucleophile approaches the C–O bond that is to be broken from behind so that its electrons can overlap with the C–O σ* orbital. You can consider this displacement an example of an S_N2 reaction where the normally unreactive ether bond is broken so as to relieve strain present in the three-membered ring.

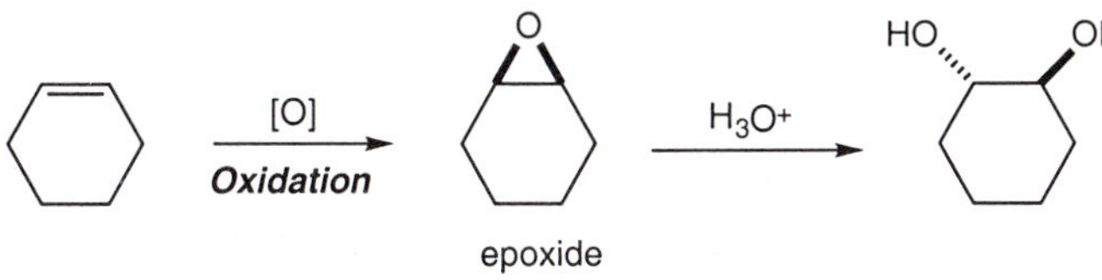

Fig. 3.1

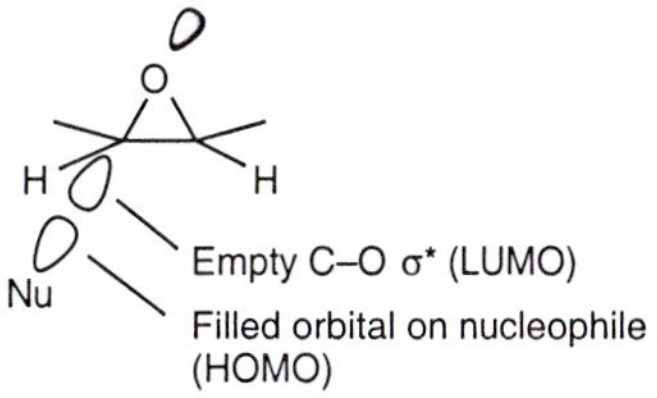

The most useful reagent for effecting the alkene–epoxide oxidation is *meta*-hloroperbenzoic acid, *m*-CPBA. The reaction is thought to proceed *via* a ingle (concerted) step which involves nucleophilic attack of the alkene π-lectrons onto the peracid, Fig. 3.2. Evidence in favour of this mechanism entres around the observation that epoxidation of alkenes with *m*–CPBA is tereospecific (see the oxidation of *E*- and *Z*-**1**). Remember that even though e oxidations shown in this figure form single diastereoisomers, the epoxide roducts are mixtures of enantiomers.

A stereospecific reaction is one in which the stereoisomers of a given starting material each give rise to different stereoisomeric products.

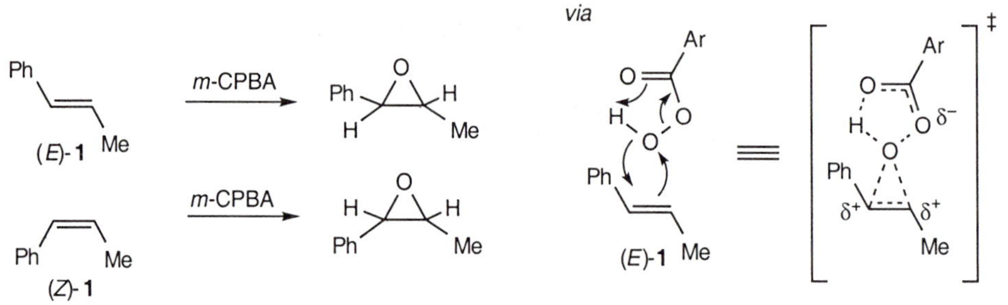

Fig. 3.2

A study of the rate of alkene epoxidation with the related peracid CH_3COOOH has shown the following relative rates: $CH_2{=}CH_2$ (1); $RCH{=}CH_2$ (24); $RCH{=}CHR$ and $R_2C{=}CH_2$ (500); $R_2C{=}CHR$ (6500); $CR_2{=}CR_2$ (>6500).

As we have just seen, *m*-CPBA is an electrophilic reagent anc accordingly, reacts well with nucleophilic alkenes. This means that the mor alkyl groups that are substituted on the alkene the better (the extra steric bul introduced by an alkyl group is orthogonal to the direction of approach of th oxidant and clearly does not do much to slow the reaction down). Using thi facet of peracid chemistry it is possible to oxidise more-substituted alkenes i the presence of less-substituted ones, so long as we do not use more than equivalent of oxidant, Fig. 3.3.

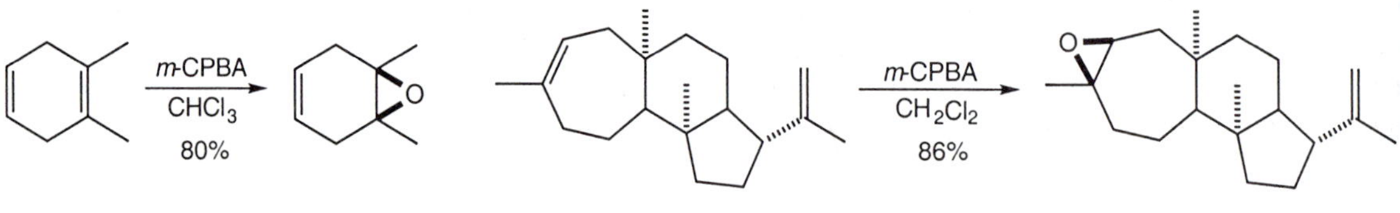

Fig. 3.3

Accordingly, alkenes that are conjugated with a carbonyl group ar epoxidised much more slowly (as they are electron-deficient) and eithe forcing conditions (*m*-CPBA at high temperatures) or use of the mor electrophilic reagent CF_3COOOH as an alternative oxidant is recommended.

A recently developed oxidant is dimethyldioxirane (DMDO) which is capable of oxidising alkenes to the corresponding epoxides in high yield. This (electrophilic) reagent is at its best when forming sensitive epoxides that are not stable to other oxidising conditions.

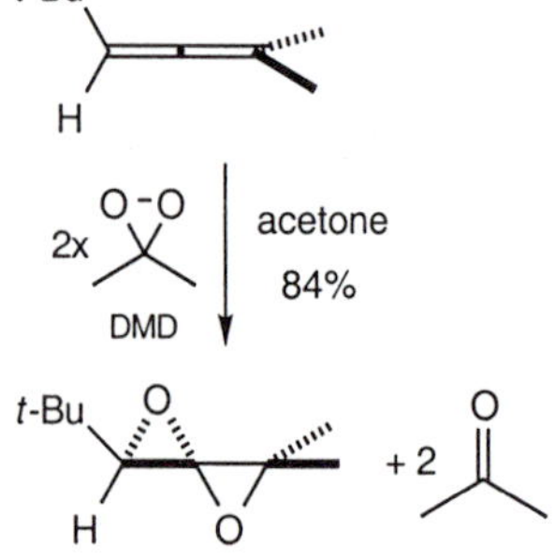

Stereoselectivity in the epoxidation reaction

Now consider an alkene unit in a molecule that is chiral— the two faces c the double bond are non-equivalent (diastereotopic) and the two possibl epoxides formed from oxidation of the alkene are diastereoisomers. In cycli systems, *m*–CPBA and other peracids tend to give the epoxide derived fror attack on the less–hindered face of the alkene, see **2**, Fig. 3.4.

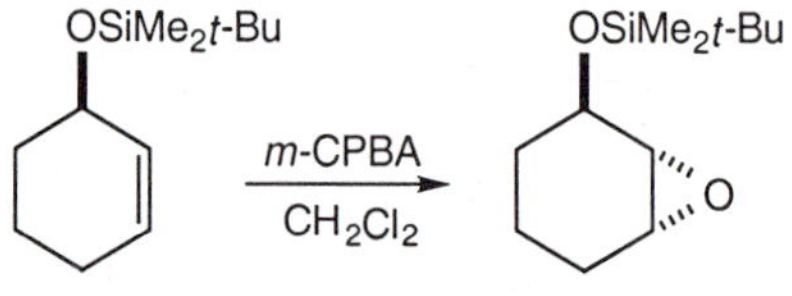

Fig. 3.4

Interestingly, oxidation of alkenes with allylic (or homoallylic) grour that are capable of hydrogen bonding (such as OH or NHAc) reveals ifferent picture. In these examples, oxidation with *m*-CPBA leads to the nost hindered epoxide, Fig. 3.5. It seems that hydrogen bonding between the cidic hydrogen on the functional group and the peracid stabilises approach of he oxidant on one face of the alkene but not the other. Compare the xidation of **2** (Fig. 3.4) with that of **3**, Fig. 3.5.

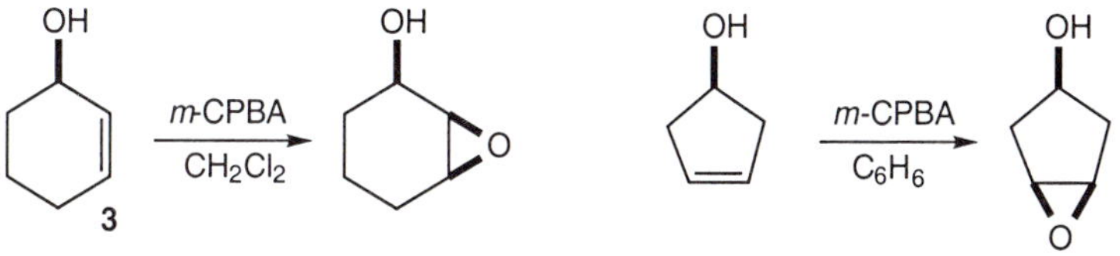

Fig. 3.5

Approach of the peracid to one face of **3** is stabilised by hydrogen bonding

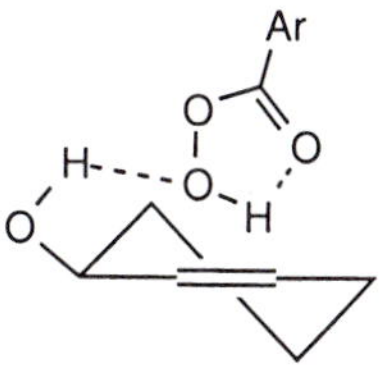

As mentioned earlier, epoxidation of electron-deficient alkenes is slow with eracids and, moreover, unsaturated ketones can react with peracids in a ifferent way (see the Baeyer–Villiger reaction, Chapter 5). An alternative pproach to oxidisng such electron-poor substrates would be to use a *ucleophilic* oxidant. Alkaline solutions of hydrogen peroxide suit this role icely. Under basic conditions, hydrogen peroxide (pKa ≈ 12) is deprotonated nd the anion (HOO^-) is a very good (soft) nucleophile towards electron-eficient alkenes, Fig. 3.6. The enolate that results from attack by HOO^- can ow act as a nucleophile towards the peroxide, breaking the weak O–O bond nd displacing hydroxide anion.

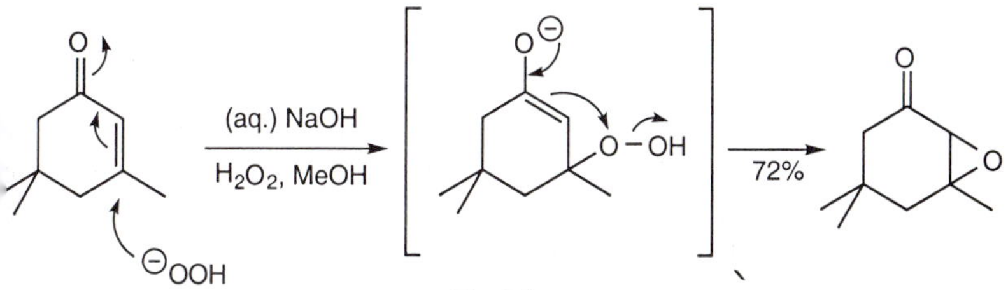

Fig. 3.6

Beware when using these oxidising conditions on acyclic enones; the action is not stereospecific, as we saw with *m*-CPBA. Instead, the reaction now a stereoselective one and both isomers of enone give predominantly ne isomer of epoxide. The crossover reaction whereby the *Z*-alkene gives the *ans* epoxide may take place by addition of peroxide anion and subsequent ond rotation before closure to the three membered ring, Fig. 3.7.

A stereoselective reaction is one which results in the preferential formation of one stereoisomer over another.

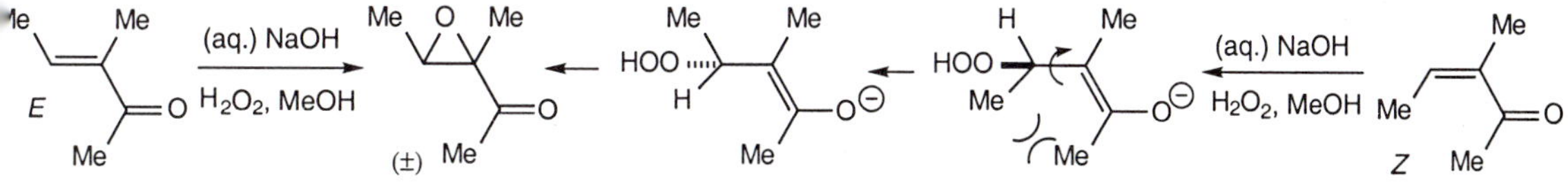

Fig. 3.7

ransition metal catalysed epoxidation of alkenes

he use of transition metals in conjunction with organic peroxides has elded a very useful set of oxidising agents (see for example the oxidation of amines and sulfides, Chapter 2). These oxidants show amazing selectivity fc the oxidation of alkenes with an allylic hydroxyl group. So, for an example consider the oxidation of geraniol which contains two trisubstituted alken units. These alkenes are not, however, identical as the C-2,3 alkene i electron deficient compared to the C-6,7 alkene (the former alkene has a allylic oxygen group withdrawing electrons inductively through the sigm system). The combination of *t*-butyl hydroperoxide and catalytic amounts c vanadium complexes ($VO(acac)_2$ is best) leads to the formation of just on epoxide, that derived from epoxidation of the C-2,3 alkene, as shown in Fig 3.8.

The reason why the vanadium-catalysed epoxidation is so specific for allylic alcohols is that a complex is formed between the hydroxyl group, transition metal and hydroperoxide; therefore the oxidant is nicely placed to epoxidise the adjacent double bond. Note that the peroxide ligand is bound to the metal in a bidentate fashion: this is so that the by-product from the oxidation, *t*-butyl alcohol, can be stabilised by coordination to the metal.

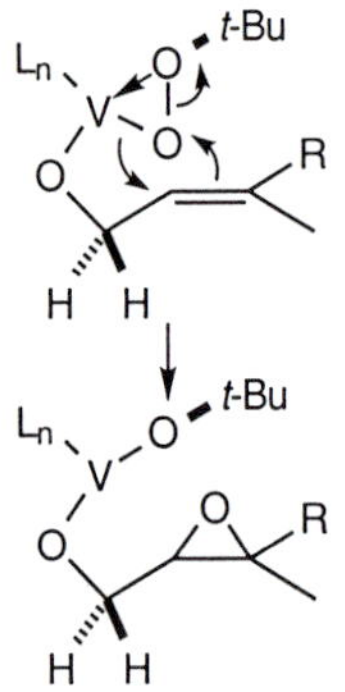

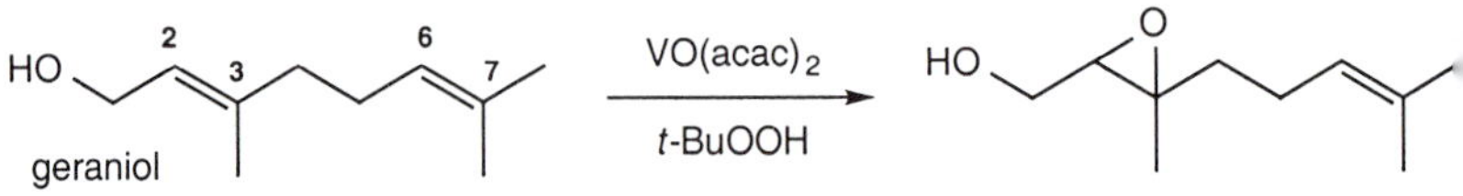

geraniol — $VO(acac)_2$, *t*-BuOOH

Fig. 3.8

The stage is now set for introduction of another level of control— that c enantioselectivity. If we use the principal ingredients from the proces described above (*t*-butyl hydroperoxide and a transition metal catalyst) i conjunction with a chiral ligand, then a *chiral* oxidant is formed, Fig. 3.9 Titanium tetraisopropoxide is the most effective transition metal complex an diethyltartrate (DET) is the best chiral ligand (DET is readily available a either enantiomer). The reaction that ensues between an allylic alcohol an this oxidant is similar to that described for vanadium— with one importa difference. The chiral ligand is able to enforce the formation of essentially single *enantiomer* of epoxide product. This reaction is an extremely usef way of introducing absolute stereochemistry into a synthetic sequence and i known as the Sharpless asymmetric epoxidation.

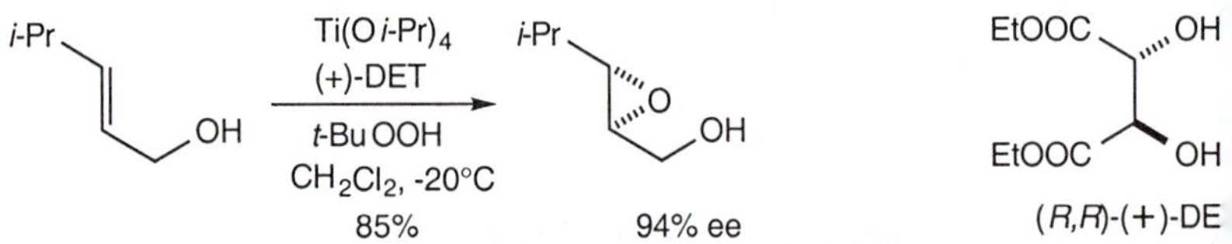

$Ti(O\textit{i}\text{-}Pr)_4$, (+)-DET, *t*-BuOOH, CH_2Cl_2, -20°C, 85% — 94% ee — (*R,R*)-(+)-DE

Fig. 3.9

Alkenes with all sorts of substitution patterns will work well in th epoxidation, although 1,2-*cis*-disubstituted alkenes do not always give hig enantioselectivity. The reaction is particularly useful because it uses catalyt quantities of transition metal— a typical stoichiometry for this reaction i $Ti(O\textit{i}\text{-}Pr)_4$ (0.1 eq.); *t*-BuOOH (2 eq.); DET (0.12 eq.); allylic alcohol (1 eq.)

The mechanism of the reaction has been studied in detail and is qui complicated. The salient features of the reaction are that the metal, ligan oxidant, and substrate form a complex which is the direct precursor to th epoxide, just as we saw with the vanadium–catalysed oxidation. Of course th fact that the alcohol complexes to both the metal and the oxidant is th reason that only allylic alcohols are epoxidised.

The structure of the active complex is almost certainly a dimer in which each titanium atom behaves independently. So, let's consider one metal atom only to rationalise the stereoselectivity, Fig. 3.10. Look at the allylic alcohol complexed to the titanium on the right–hand side of the dimer; the alkene adopts a conformation whereby it can lie above the active oxygen of the peroxide. As we have seen before, the peroxide is bound in a bidentate mode to the transition metal. The transition structure **A** (which leads to the formation of one enantiomer) is favoured over the structure **B** (which would lead to the other enantiomer). This is because of unfavourable steric interactions in **B** between the allylic alcohol and the ester group (E) of the ligand and also with the *t*-butyl group of the peroxide.

The Sharpless epoxidation is so reliable that a general rule has been developed to help chemists decide which enantiomer of DET will produce which epoxide. Draw the allylic alcohol in the orientation illustrated and (-)-DET will epoxidise it from the top face and (+)- DET from the lower face.

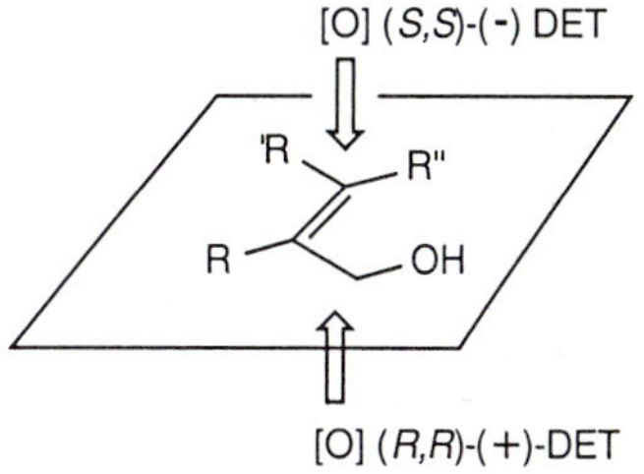

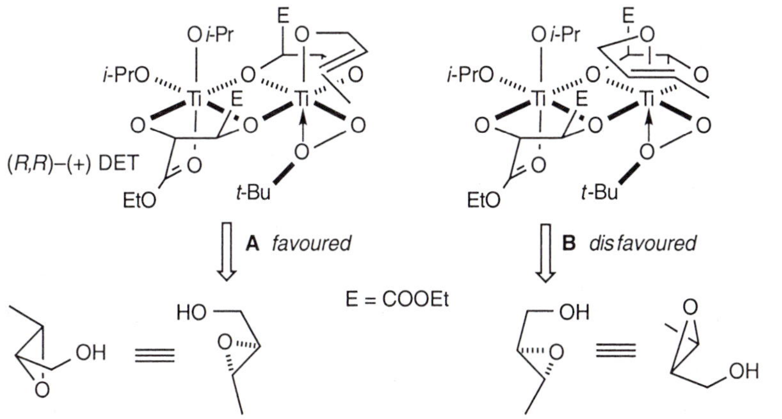

Fig. 3.10

You can see from what has been presented so far in this section that the epoxidation of most types of alkenes can be accomplished by the judicious choice of oxidant, and that this general reaction has reached a very sophisticated stage.

3.2 Addition to alkenes *via* epi-ions

You should have come across the reaction between an alkene and a halogen (Br_2, Cl_2) to give a vicinal dihalide. Perhaps the most useful member of the halogens for this purpose is bromine, which adds readily to most alkenes, Fig. 3.11. This reaction proceeds *via* the intermediacy of an epi-ion (called a bromonium ion in this case) which is opened in a second step by reaction with bromide ion. Opening of the epi-ion takes place with inversion of configuration (just like the opening of epoxide that we saw earlier). This means that the overall reaction is stereospecific.

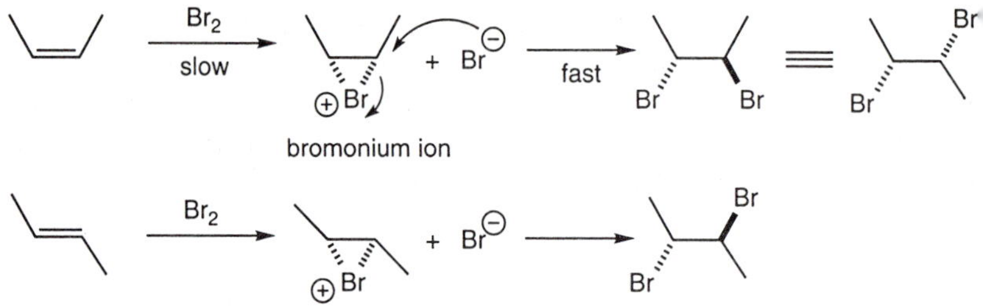

Fig. 3.11

We can prove the existence of bromonium ions! The bromonium ion shown below has been prepared from the corresponding alkene and is quite stable. This species cannot be attacked easily by a nucleophile due to steric hindrance around the back of the C–Br bonds. The Br_3^- counterion is a product of the addition of Br^- to Br_2.

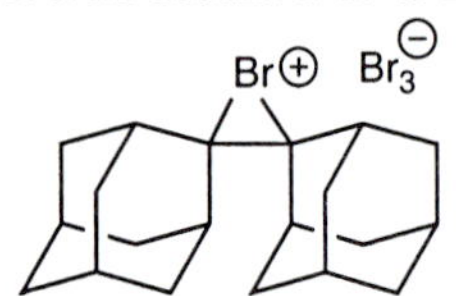

Formation of the bromonium ion is rate determining in these reactions and in this step the alkene acts as a nucleophile towards electrophilic bromine. Consequently, as an alkene becomes more substituted (and therefore more electron rich) it is brominated more rapidly, just like epoxidation. Approximate relative rates of bromination are CH_2=CH_2 (1); RCH=CH_2 (61); RCH=CHR (2000); R_2C=CH_2 (5400); R_2C=CHR (1.3 × 10^4); CR_2=CR_2 (1.8 × 10^5).

Of course, if we add a halogen to an alkene in the presence of an external nucleophile (usually in the form of the solvent) then another pathway can be followed as the epi-ion is opened by the additional nucleophile (if this is solvent then it will be present in a much greater concentration than X^- which cannot compete as a nucleophile). Figure 3.12 illustrates this point with iodine as the halogen and methanol as both solvent and nucleophile. Just like the bromination reaction which preceded it in this section, the reaction is a stereospecific one.

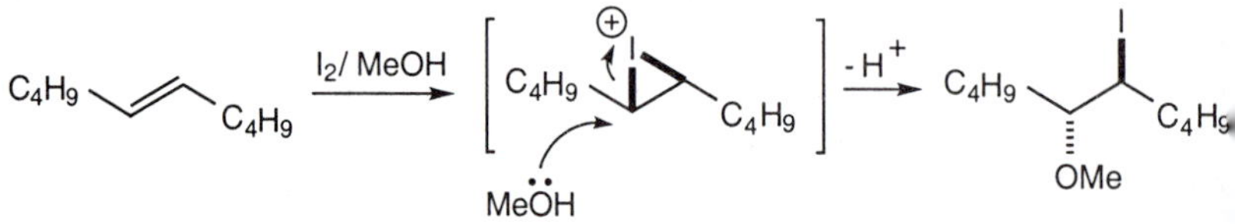

Fig. 3.12

Oxidation of an alkene with iodine and silver benzoate (PhCOOAg abbreviated to AgOBz) yields the corresponding *bis*-benzoate, Fig. 3.13. This sequence, known as the Prevost reaction, proceeds *via* a β-iodobenzoate formed from an iodonium-ion which reacted with benzoate ion. This product reacts further with excess silver benzoate (the Ag^+ cation helps to make iodide into a better leaving group) to give the final product. The stereochemistry of this product tells us that the displacement of iodine occurred with retention of configuration and double inversion is the culprit!

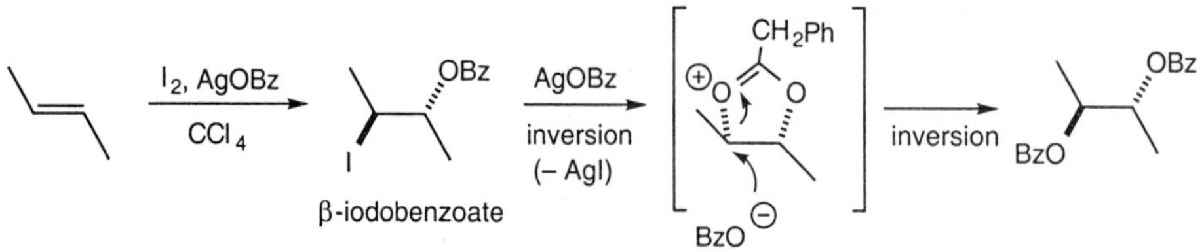

Fig. 3.13

Woodward made modifications to the Prevost reaction such that iodine and silver acetate in wet acetic acid were the reagents. The result was formation of the monoacetate derivative of a *syn* 1,2–diol, Fig. 3.14. An intermediate common to the Prevost reaction seems likely (with CH_3 replacing $PhCH_2$) and water attacks the central carbon to form the *syn* diol derivative.

Note the different stereochemistry that the Prevost and the Woodward reactions give on the same substrate.

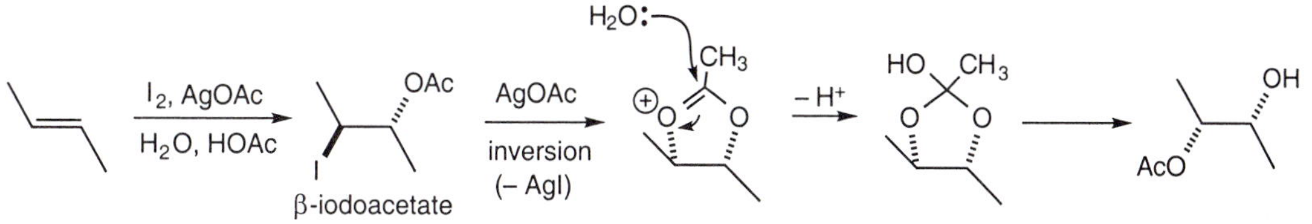

Fig. 3.14

We will mention in passing here some other oxidation reactions that proceed (i) *via* an epi-ion and (ii) result in the introduction of oxygen, so as to conform to our definition of oxidation given earlier. PhSeCl reacts with alkenes, in manner reminiscent of Br_2 described earlier. In these cases, the alkene attacks the more electropositive end of the electrophile (*ie*. selenium rather than chlorine) to form an epi-ion which can be attacked either by chloride ion or by an external nucleophile, Fig. 3.15. The stereochemistry of the product shows that the nucleophilic attack took place with inversion. As you might imagine, this generalised sequence gives access to a wide variety of alkanes substituted with two adjacent heteroatoms.

One issue that we have so far avoided is that of regiochemistry; i.e. which carbon of the unsymmetrical epi-ion **4** will be attacked by a nucleophile? Fortunately, a general rule can be formulated which states that the nucleophile will attack the *most* substituted end of the epi-ion. At first sight this may seem counterintuitive as the greater substitution at this carbon would be expected to slow down the rate of nucleophilic attack. However, one must remember that the most substituted end of the epi-ion can sustain the greater positive charge and is therefore the most attractive to nucleophiles in a S_N1 like reaction.

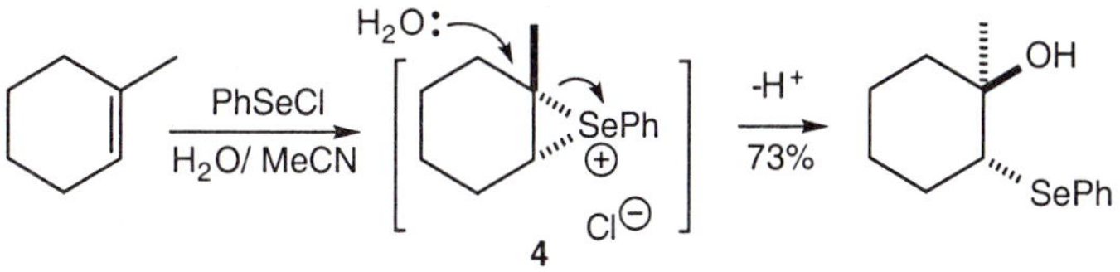

Fig. 3.15

3.3 Oxidation of alkenes to *syn* diols

There is only one really reliable method of oxidising alkenes to form *syn*-1,2-diols and that involves osmium tetroxide (OsO_4) as an oxidant. Both oxygens in the diol derive from one molecule of osmium tetroxide and these are delivered to the alkene in essentially one step. This addition is another classic example of a stereospecific oxidation: stereochemistry in the alkene starting material (*cis* or *trans*) is transferred to stereochemistry in the product, Fig.

3.16. Contrast the *syn*–diol product **5** with the *anti*–diol product **6** one would obtain from the same alkene by epoxidation and subsequent hydrolysis with aqueous acid.

It has been found that more substituted alkenes react fastest with osmium tetroxide thus leading to the notion that the oxidant plays an electrophilic role in the reaction. Approximate relative rates of reaction are RCH=CH_2 (1); RCH=CHR (1.2–3); RRC=CHR (5); R_2C=CR_2 (30).

Osmium tetroxide is quite expensive and so conditions were developed that enabled it to be used as a catalyst: workers at the Upjohn company found that 5 mol% osmium tetroxide was sufficient to oxidise an alkene if a stoichiometric amount of *N*-methylmorpholine-*N*-oxide (NMO) was used to reoxidise the transition metal *in situ*. This reaction works rather well and so has become known colloquially as the Upjohn reaction.

OsO_4 (cat.)
NMO (1.5 eq.)
acetone/ water
m-CPBA
H_3O^+
OH HO **5**
6 HO OH

Fig. 3.16

Some years ago it was noted that an amine additive speeds up the dihydroxylation reaction by about one to two orders of magnitude: this effect is related to the fact that amines coordinate to osmium tetroxide and form (18–electron) complexes. Several amines have been examined in the role of accelerating ligand and quinuclidine is one of the best for speeding up an otherwise slow reaction.

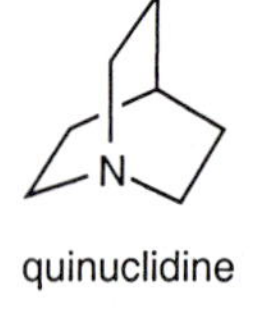

quinuclidine

The exact mechanism of osmium tetroxide oxidation is a contentious one. Two schools of thought exist, one of which advocates a [3+2] addition of OsO_4 to the alkene in one step— this leads directly to an osmate ester which is hydrolysed (and oxidised) to liberate the diol, Fig. 3.17. Another possible mechanism involves the addition of the alkene across the osmium oxygen double bond to form an unstable osmaoxetane intermediate: this may rearrange to the same osmate ester as proposed for the [3+2] mechanism. Recent studies have shown that both mechanisms may be operative, depending on the exact reaction conditions employed.

[3+2] OsO_4
osmate ester
oxidation (NMO) and hydrolysis
+ OsO_4
AND/OR
[2+2] OsO_4
osmaoxetane
rearrangement
oxidation and hydrolysis as above

Fig. 3.17

Stereoselectivity in the dihydroxylation reaction

Osmium tetroxide is a rather bulky reagent and therefore it oxidises chiral alkenes from the most sterically accessible face, with good levels of stereoselectivity, Fig. 3.18. When oxidising allylic alcohols, osmium tetroxide does not normally hydrogen bond to the acidic hydrogen (although exceptions are known).

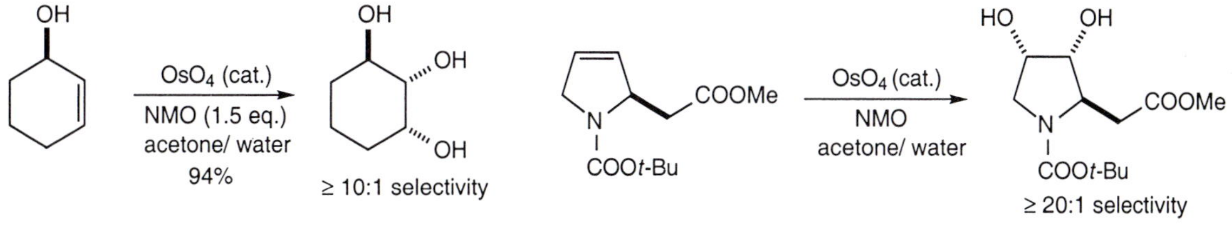

OsO_4 (cat.), NMO (1.5 eq.), acetone/ water, 94%; ≥ 10:1 selectivity

OsO_4 (cat.), NMO, acetone/ water; ≥ 20:1 selectivity

Fig. 3.18

Sharpless has recently developed conditions which lead to control of the *enantioselectivity* of the dihydroxylation reaction. The accelerating effect of an amine ligand during the dihydroxylation reaction is crucial to the success of Sharpless' chemistry. Use of a *chiral* amine enabled osmium tetroxide to add to one face of the alkene over the other, leading to diol products with high levels of enantioselectivity, Fig 3.19. The best chiral amines are based on two alkaloids, dihydroquinidine (DHQD), which produces one enantiomer of the diol product, and dihydroquinine (DHQ), which gives the other: although these amines are not enantiomers, their structures are very nearly mirror images of each other and, to all intents and purposes, they behave as enantiomers in the dihydroxylation reaction. Both DHQD and DHQ are obtained from natural sources and are readily available.

DiHydroQuinine (DHQ) DiHydroQuiniDine (DHQD)

Fig. 3.19

The Sharpless asymmetric dihydroxylation is even more remarkable in that it only uses catalytic amounts of osmium tetroxide and chiral ligand. In most cases NMO acts as a competent re-oxidant for osmium, although K_3FeCN_6 is also used.

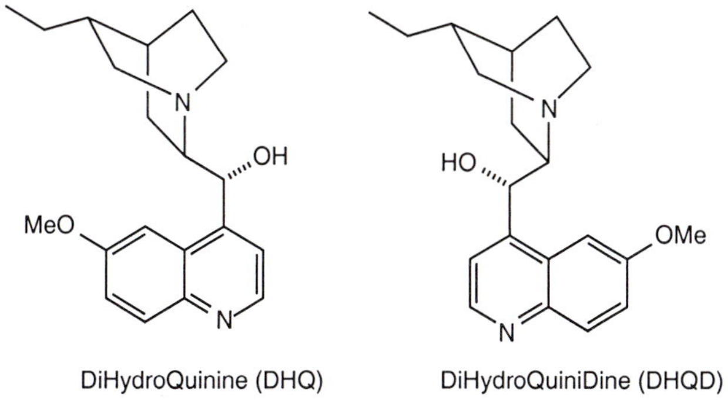

In fact, the only complication with this chemistry is that there is no one general chiral amine catalyst that is specific for all types of alkene, and one must choose from a selection. Each of the most active catalysts is a dimer of either DHQ or DHQD joined (at the hydroxyl group of the alkaloid) by a spacer group. One must select the spacer group that gives rise to the most active catalyst for a particular alkene substitution pattern.

Two of the most common spacer groups are illustrated below, Fig. 3.20 (both are joined to DHQ alkaloids although they could just as easily be joined to DHQD ligands).

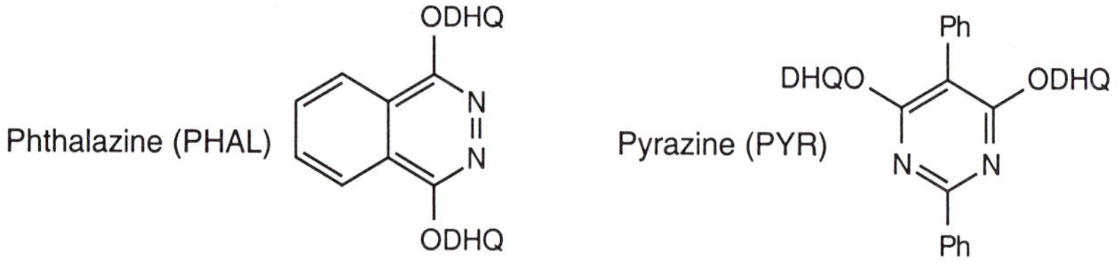
ODHQ
Phthalazine (PHAL)
N
N
ODHQ
Ph
DHQO
ODHQ
Pyrazine (PYR)
N
N
Ph

Fig. 3.20

Many alkenes have been oxidised using these conditions and generally diol products with very high enantiomeric excesses are obtained. For example, mono-, 1,1-di- and *trans*-disubstituted alkenes give diols with 90–99% ee. Perhaps the least useful class of alkenes for this chemistry are *cis*-disubstituted alkenes which lead to poor (40–60% ee) levels of stereoselectivity. However, most trisubstituted and some tetrasubstituted alkenes are also good substrates and give products with very high enantiomeric purities.

The oxidation of (*E*)-stilbene to the corresponding diol can be accomplished with OsO_4, NMO and $(DHQ)_2$PHAL to give the diol shown in ≥ 99% ee. This reaction can be performed easily on a kilogram scale.

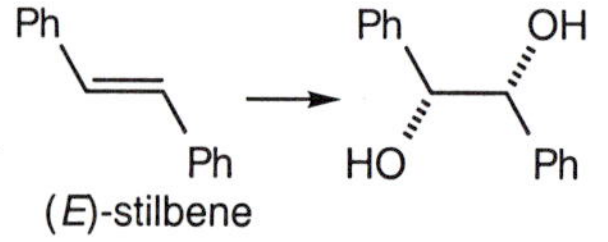
Ph
Ph
(*E*)-stilbene
Ph
OH
HO
Ph

The origins of the stereoselectivity that these chiral catalysts enforce are not confirmed, but it is clear that the (chiral) amine binds to osmium tetroxide *via* its *tertiary* amine and in so doing enables the oxidant to discriminate between the two enantiotopic faces of the alkene. Remember also that the chiral amine speeds up the reaction considerably; this means that the background reaction of osmium tetroxide on its own (which would lead to racemic products) is not competitive with the enantioselective one. The general phenomenon whereby a ligand accelerates a catalytic reaction is extremely useful in organic synthesis.

The asymmetric dihydroxylation reaction has been used to good effect in some spectacular reactions, which illustrate the high levels of control that are possible, Fig. 3.21.

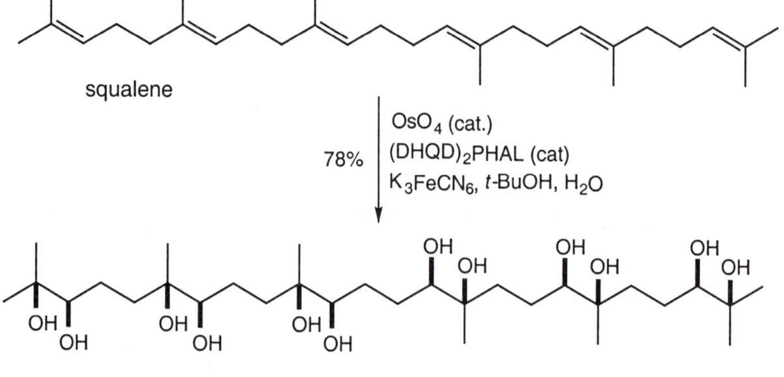
squalene
78%
OsO_4 (cat.)
$(DHQD)_2$PHAL (cat)
K_3FeCN_6, *t*-BuOH, H_2O
OH
OH
OH
OH
OH
OH
OH
OH
OH
OH
OH
OH

one diastereoisomer AND one enantiomer, after recrystallisation of an acetonide derivative

Fig. 3.21

Singlet oxygen oxidation of dienes to 1,4–diols

Oxygen that is present in air is a triplet diradical that participates in radical reactions with organic compounds. However, irradiation of triplet oxygen in the presence of a sensitizer generates singlet oxygen which behaves as O=O. So, [4+2] cycloaddition of singlet oxygen with dienes generates a peroxide that is usually reduced to form a *cis*–1,4–diol, Fig. 3.22.

A sensitizer is a compound that effectively absorbs the incipient light and is excited. Its excited state then transfers energy to triplet oxygen, so forming the singlet state, $^1O_2$.

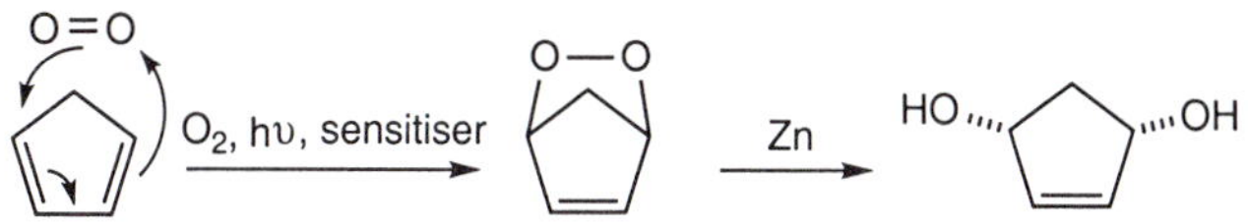

O=O; O_2, hυ, sensitiser; O—O; Zn; HO, OH

Fig. 3.22

The peroxide cleavage proceeds according to the general equation: $ROOR + Zn \rightarrow 2\ RO^- + Zn^{2+}$

3.4 Formation of ketones from alkenes: the Wacker process

The oxidation of alkenes to form ketones can be achieved by reaction with palladium(II) salts and oxygen; the name of the reaction derives from the company that developed the prototype reaction, that of ethylene to acetaldehyde. The sequence involves addition of $PdCl_2$ to an alkene in water, Fig. 3.23.

While terminal alkenes give predominantly the ketone isomer rather than the aldehyde product, such control of regiochemistry is not always easy to achieve with 1,2-disubstituted alkenes.

R; $PdCl_2$, H_2O; $CuCl_2$, O_2; R; O; R; R'; $PdCl_2$, H_2O; $CuCl_2$, O_2; R; R'; O; +; O; R; R'

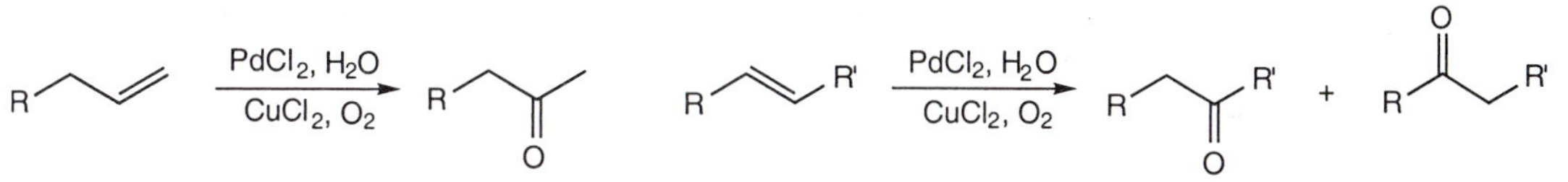

Fig. 3.23

Moreover, increasing substitution on the alkene seems to retard the rate of oxidation (1,2-disubstituted alkenes are approximately five times slower in the reaction than terminal alkenes). This difference in rate can be used for selective oxidation in synthesis; compound **7** was subsequently used in a synthesis of muscone, Fig. 3.24.

Muscone is used in the perfumery industry and its natural source is the musk deer. Synthesis seems to offer a much better route to this compound!

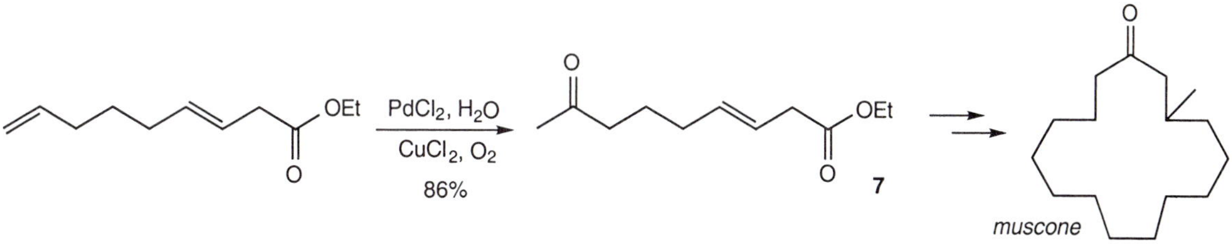

OEt; O; $PdCl_2$, H_2O; $CuCl_2$, O_2; 86%; O; OEt; O; **7**; O; *muscone*

Fig. 3.24

Re-oxidation of Pd(0) is achieved by the addition of Cu(II) salts and by performing the reaction in the presence of air. The following generalised reaction is thought to ensue, Pd(0) + 2Cu(II) → Pd(II) + 2Cu(I). Fortunately, Cu(I) is then re-oxidised itself back to Cu(II) by oxygen, so continuing the cycle.

As the reaction proceeds, and the alkene is oxidised, palladium(II) is reduced to palladium(0). As palladium is quite expensive, a method was developed for the *in-situ* oxidation of Pd(0) back to Pd(II) so that only catalytic amounts of the transition metal are required.

The mechanism of the Wacker process is a matter for debate, but it is interesting to note that deuterium–labelling studies on the ethylene to acetaldehyde reaction have shown that all four hydrogen atoms in the product originate from the starting alkene and not from the solvent. The following figure shows the key parts of one mechanism that is consistent with the labelling work, Fig. 3.25.

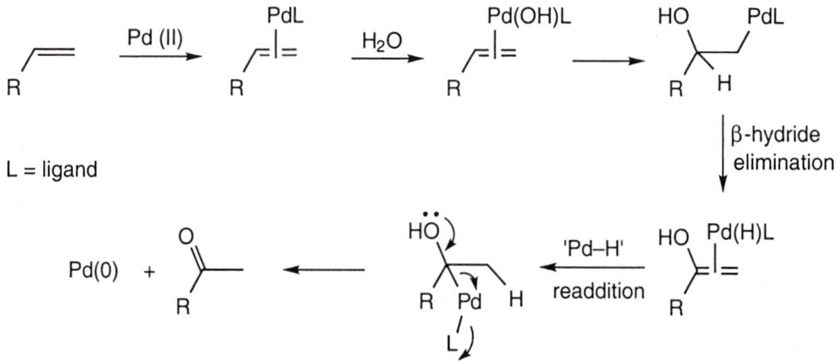

Pd (II) → PdL; H_2O → Pd(OH)L; → HO PdL R H; β-hydride elimination; HO Pd(H)L R; 'Pd–H' readdition; HÖ R Pd H L; Pd(0) + O R

L = ligand

Fig. 3.25

3.4 The oxidation of alkynes to 1,2-diketones

RuO_2 is not the active oxidant in this reaction; rather it is oxidised by periodate to RuO_4 which is a powerful oxidising agent often generated *in situ.* We will come across this reagent combination later in the book.

The π-bonds of alkynes are reasonably susceptible to oxidation, and give products rather like those from alkenes. However, alkynes are in a higher oxidation level than alkenes and the products are going to be more highly oxidised than those from the corresponding alkene. A useful transformation is that of an alkyne into a 1,2-diketone and, while a variety of oxidants (O_3, $KMnO_4$) will accomplish this task, the mixture of RuO_2 (catalytic) and $NaIO_4$ appears to be the most general, Fig. 3.26.

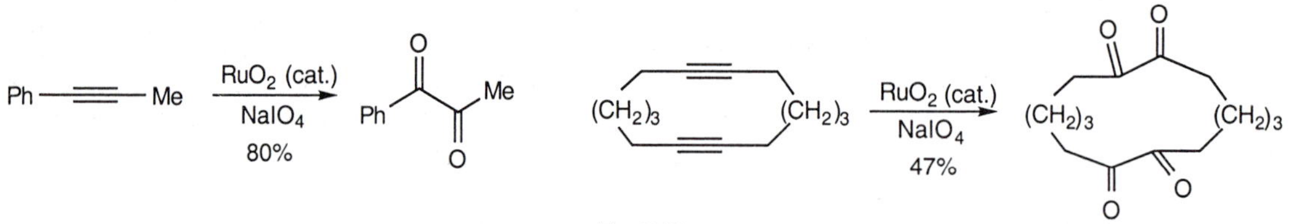

Ph—≡—Me → RuO_2 (cat.), $NaIO_4$, 80% → Ph(C=O)(C=O)Me

$(CH_2)_3$ diyne macrocycle → RuO_2 (cat.), $NaIO_4$, 47% → tetraketone $(CH_2)_3$

Fig. 3.26

4. Oxidation of activated carbon–hydrogen bonds

n previous chapters we have not really stretched the definition of oxidation ully, as we have mainly considered reaction of polarisable electrons (such as one pairs on heteroatoms or alkene systems) with electrophilic oxidising gents, so increasing the oxygen content of the substrate. What about emoval of hydrogen from an organic compound— surely this definition of xidation has some useful manifestations in organic synthesis? During the ourse of an oxidation, hydrogen can be removed in three ways, as either H^+, $I^\bullet$, or H^-; we shall cover examples of these processes in this chapter. Iormally, it is beneficial to have a functional group which stabilises the arbon left behind by such removal of hydrogen; so I consider the hydrogen tom that is lost (in whatever guise) activated by the functional group.

.1 Oxidation adjacent to oxygen

'his is the most useful and widespread area of oxidation in this chapter, as it ncompasses the alcohol → aldehyde → carboxylic acid sequence that is used o often in synthesis.

)xidation of alcohols to aldehydes and ketones

'onsider the transformation of an alcohol into a carbonyl group; this reaction s loss of hydrogen (dehydrogenation) of an alcohol. And, as befits such an nportant reaction, there are many reagents which will achieve this oxidation: his is a good thing as it means that there are always plenty of oxidants to try vhen oxidising an unstable or unreactive molecule.

Let's begin with chromic acid (H_2CrO_4 which is made *in situ* by the eaction of H_2O and CrO_3). This reagent is normally used in conjunction with ulfuric acid and as such this mixture is a powerful oxidant, Fig. 4.1.

The combination of CrO_3, H_2SO_4 and acetone solvent is called the Jones reagent. Functional groups such as alkenes, which are acid-sensitive and potentially oxidisable, are usually left untouched under these conditions.

Primary alcohols are oxidised to aldehydes by the Jones reagent and econdary alcohols are oxidised to ketones. Yields are variable for the first ransformation because over oxidation of an aldehyde to the corresponding arboxylic acid (RCOOH) normally occurs *in situ*. In fact it is much more ommon to use the Jones reagent to transform primary alcohols directly into arboxylic acids.

In general it is easier to stop the Jones oxidation of primary alcohols at the aldehyde stage if the R group in Fig. 4.1 is unsaturated.

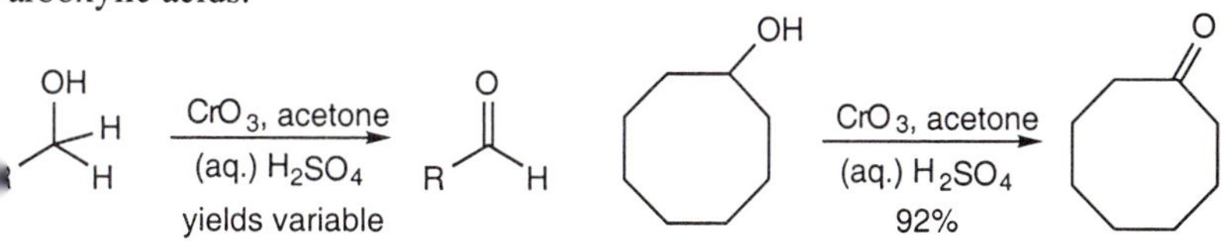

CrO_3, acetone / (aq.) H_2SO_4 / yields variable

CrO_3, acetone / (aq.) H_2SO_4 / 92%

Fig. 4.1

The mechanism of this oxidation is thought to entail the initial formation of a complex between the alcohol and the chromium(VI) reagent: this is known as a chromate ester, Fig. 4.2. The second stage of oxidation involves a breakdown of this complex to liberate a carbonyl compound and a reduced chromium(IV) complex.

This mechanism gives us a clue as to how aldehydes are oxidised to carboxylic acids with chromic acid. Formation of a small amount of hydrate *in situ* would give a substrate which could be oxidised to an acid *via* a chromate ester.

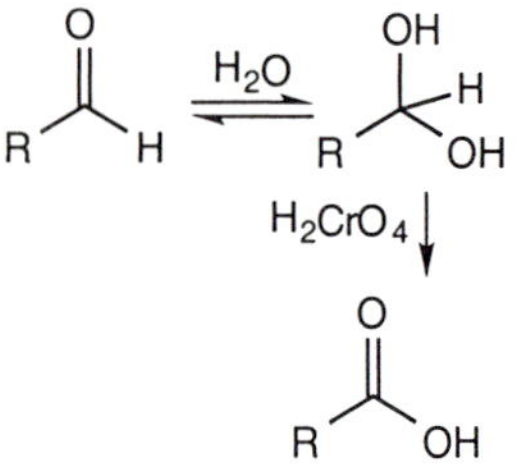

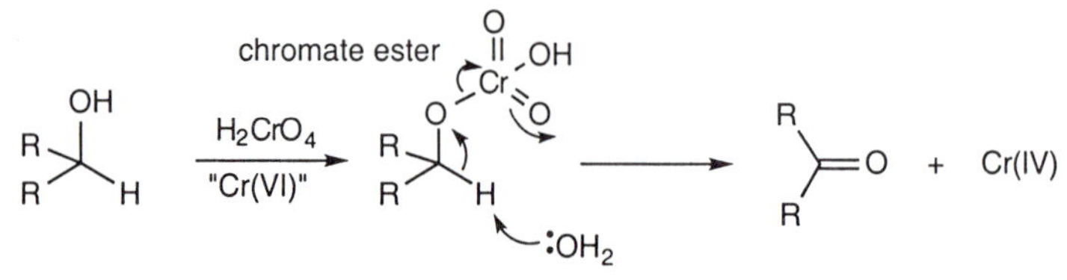

Fig. 4.2

In general, formation of the chromate ester is fast and it is the second step, the breakdown of this ester, that is rate determining. Experimentally it has been shown that more hindered secondary alcohols oxidise faster than less hindered ones: this is because hindered chromate esters collapse more quickly to relieve steric strain. Using this rationale, we can understand why, in the oxidation of secondary alcohols attached to a six-membered ring, axial alcohols (which are relatively hindered) are oxidised to ketones about three times faster than the equatorial isomers.

Oxidation of primary alcohols to aldehydes

So, the transformation of a primary alcohol into an aldehyde can be seen as a problem: how do we avoid over oxidation to a carboxylic acid?

Two solutions to this problem, based on chromium(VI) chemistry, are (i) pyridinium chlorochromate (PCC) and (ii) pyridinium dichromate (PDC). Both of these reagents (used in a non-aqueous solvent such as dichloromethane) give excellent yields of ketones from secondary alcohols and aldehydes from primary alcohols. Presumably, the success of these reagents is attributable to their compatibility with anhydrous conditions where no water is present to hydrate the aldehyde product and so lead to over oxidation.

Both PCC and PDC can be prepared in the laboratory.

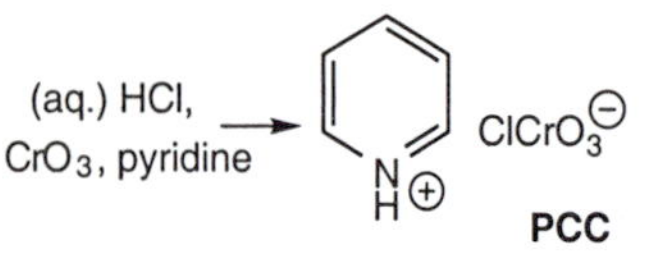

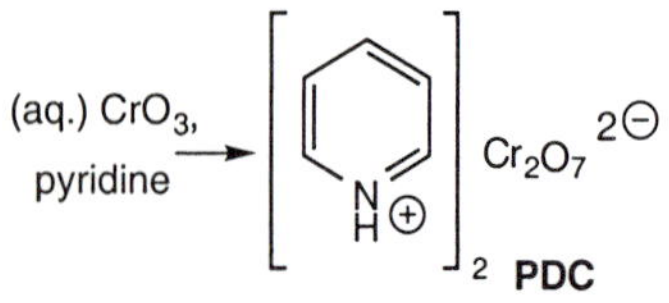

Perhaps the most general solution to the problem is the Swern reaction which involves reaction of a primary alcohol with triethylamine, dimethyl sulfoxide (DMSO) and oxalyl chloride $(COCl)_2$, all at low temperature, Fig. 4.3. This mixture is particularly adept at transforming primary alcohols into aldehydes. There are many good reasons for the popularity of this reaction and these include the use of mild conditions, volatile by-products and lack of over oxidation of aldehydes to carboxylic acids.

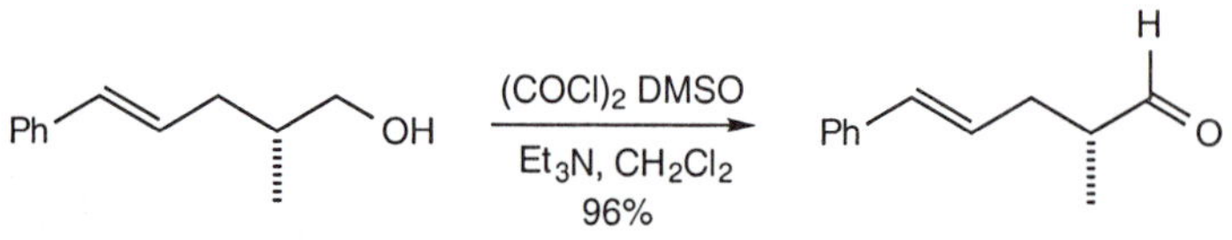

Fig. 4.3

The mechanism of the Swern oxidation involves reaction of DMSO with xalyl chloride to form an activated species **1**, Fig. 4.4. The alcohol to be xidised is then added to compound **1** and nucleophilic attack at the positively narged sulfur ensues to form **2**; this collapses to form the carbonyl ompound. The mechanism by which **2** collapses has been shown to involve itial deprotonation adjacent to sulfur (forming ylid **3**) and subsequent itramolecular proton abstraction. Note that as the alcohol is oxidised DMSO reduced to dimethyl sulfide (Me_2S).

Another advantage of the Swern oxidation is that the formation of aldehydes with α-stereogenic centres takes place without appreciable racemisation (this is something that can occur with other oxidising agents due to enolisation under acidic or basic conditions).

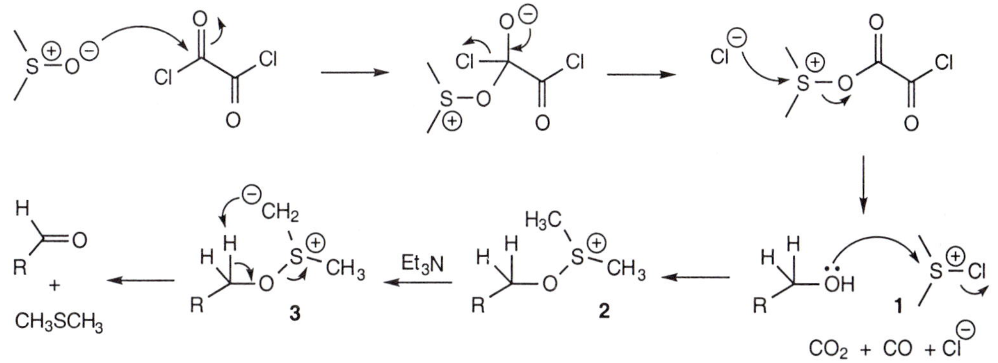

The active species in the Swern reaction, **1**, can also be prepared by the action of dimethyl sulfide with chlorine.

Fig. 4.4

An additional method of accomplishing the alcohol to aldehyde oxidation ivolves the Dess–Martin periodinane, **4** Fig. 4.5. This mild oxidant is asily prepared on a large scale from *o*-iodobenzoic acid, potassium bromate nd acetic anhydride, and is a reliable and selective reagent (for example it oes not oxidise sulfides or alkenes). Using wet CH_2Cl_2 as a solvent increases e rate by which alcohols are oxidised and was essential, for example, in the xidation of the hindered secondary alcohol **5**.

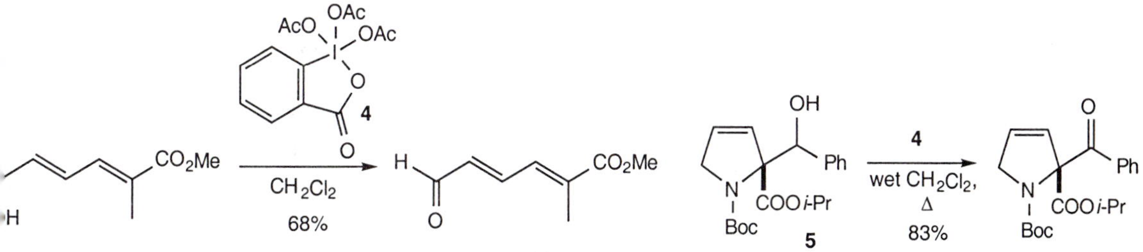

Fig. 4.5

The mechanism by which the periodinane acts is not known with ertainty, but involves initial exchange of an acetate ligand on iodine with the cohol functionality to give **6**, Fig. 4.6. This intermediate then collapses to rm the carbonyl compound, reduced iodine compound and acetic acid.

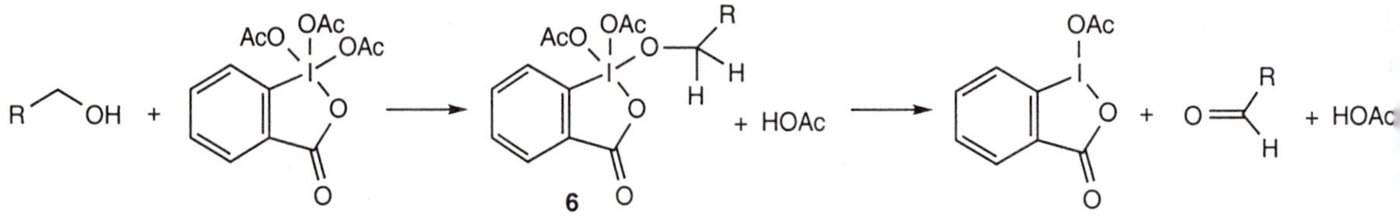

Fig. 4.6

Another convenient method for oxidising alcohols to aldehydes or ketone utilises the perruthenate ion (RuO_4^-). TPAP (tetra-*n*-propylammoniur perruthenate) is a commercially available, organic–soluble oxidant which ha been used to effect oxidation in complex molecules without disturbing othe functional groups, Fig. 4.7. Aldehydes bearing a labile α-stereogenic centr can be prepared without scrambling of stereochemistry. One of th advantages of this oxidant is that it can be used as a catalyst (typically mol%) in conjunction with an excess of NMO (used to re-oxidise th ruthenium *in situ,* compare with OsO_4). This reagent combination is efficier in large–scale synthesis as long as the reaction is kept free of moisture.

Ruthenium tetroxide is a particularly strong oxidising agent which frequently oxidises primary alcohols to carboxylic acids and cleaves alkenes. Perruthenate ion, which may be considered as a reduced form of RuO_4, is a milder and more selective oxidising agent.

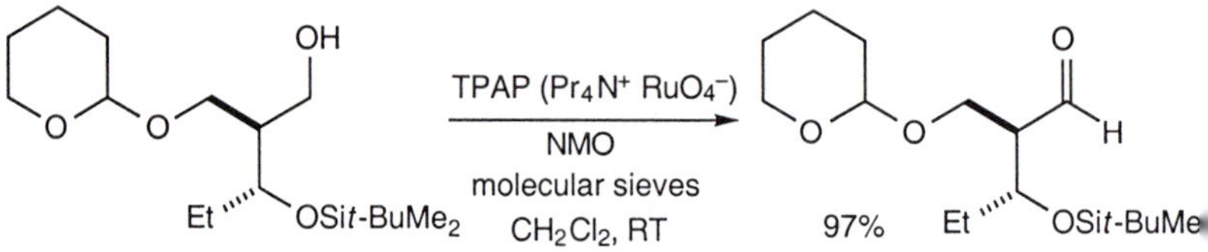

Fig. 4.7

Selective oxidation of allylic and benzylic alcohols to carbonyl groups

Manganese dioxide is a useful reagent for oxidising primary alcohols t aldehydes and secondary alcohols to ketones. However, under norma conditions MnO_2 is not sufficiently reactive to oxidise saturated alcohols an so only hydroxyl groups that are further activated by an adjacent unsaturate group (e.g. an alkene or an aromatic ring) are transformed. The resultin selectivity can be useful in organic synthesis, Fig. 4.8.

Fig. 4.8

The key to a successful oxidation with manganese dioxide is the preparation of an active form prior to use. This is best done by treatment of MnO_2 with basic potassium permanganate, followed by heating.

The reaction has been suggested to proceed *via* formation of a manganes ester complex such as **7**, followed by hydrogen atom abstraction (i.e. a radica mechanism). Finally, cleavage of the oxygen–manganese bond provides carbonyl compound and a reduced manganese(II) complex. Breakdown of could reasonably be assumed to proceed *via* a pathway similar to tha described for the collapse of chromate esters, Fig. 4.9. However, radicals hav been implicated in this particular reaction, thus leading to the modifie heme shown below: whatever the precise mechanism for this step, the tcome is the same.

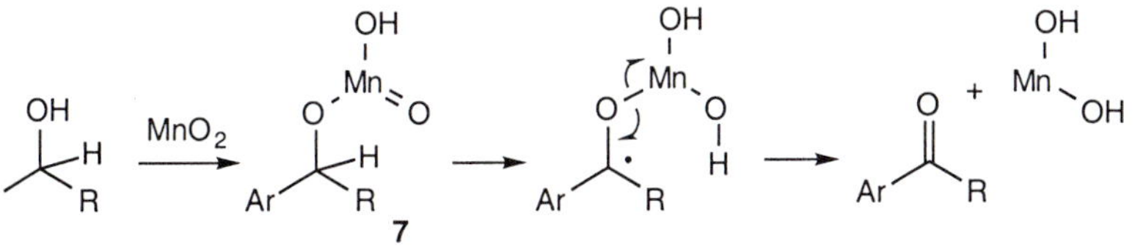

Fig. 4.9

If hydrogen atom abstraction were the rate–determining step in the oxidation, then we would expect it to be faster for allylic and benzylic alcohols as the incipient radical is stabilised by orbital overlap with the adjacent π–system. This explains the observed rate difference for such substrates.

xidation to form carboxylic acids and nitriles

e have already seen that the Jones reagent (CrO_3, H_2SO_4) is capable of idising primary alcohols directly into carboxylic acids, *via* the intermediacy an aldehyde. Therefore, it should come as no surprise to note that aldehydes emselves may be efficiently converted into carboxylic acids with a number oxidising species (H_2CrO_4, $KMnO_4$). One particularly useful oxidant that suitable for selective oxidation in multi-functional compounds is sodium lorate ($NaClO_2$), Fig. 4.10. This oxidant is inexpensive and has the tential to be useful in large-scale reactions.

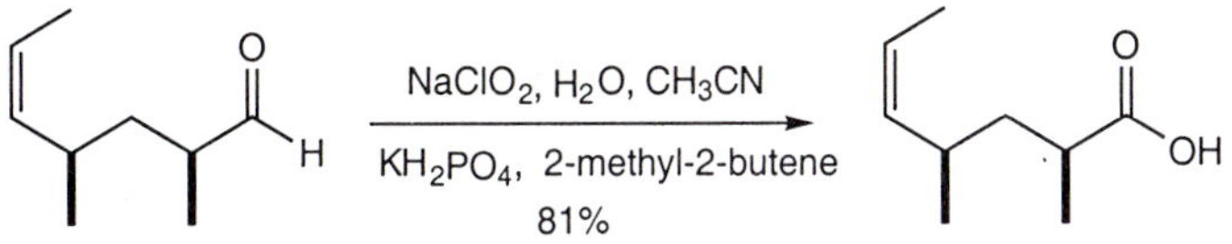

Fig. 4.10

The overall reaction proceeds according to the following equation:

$$RCHO + HClO_2 \longrightarrow RCOOH + HOCl$$

KH_2PO_4 is used as a buffer to maintain a reaction pH of 4, and 2-methyl-2-tene is used to 'mop up' the HOCl by–product from the oxidation resumably this involves HOCl acting as an electrophile and forming a loronium ion by reaction with the alkene— this may then be opened by ater in a separate step).

A neat method of converting aldehydes into nitriles (these can of course be drolysed to carboxylic acids in a separate step) involves formation of an ime by reaction with hydroxylamine and then subsequent dehydration with etic anhydride. Although this reaction is conceptually two steps (oxime rmation and elimination), both can be accomplished in a single pot by action of an aldehyde with hydroxylamine and acid, Fig. 4.11.

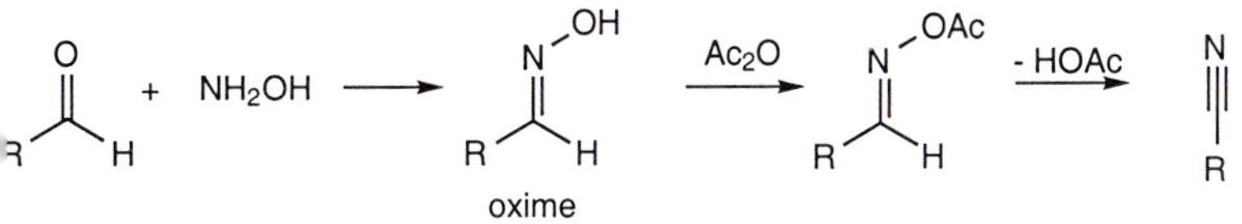

Fig. 4.11

2 Oxidation adjacent to a carbon–carbon multiple bond

xidation at allylic positions

xidation at an allylic position is a useful transformation in organic nthesis. Clearly any oxidant that we use will also have the potential to oxidise the alkene itself, and so reagents and reaction conditions must b chosen with care.

Selenium dioxide is a unique reagent that will transform alkenes in allylic alcohols in one step, Fig. 4.12. Conditions have been develop which utilise *t*-BuOOH as a re-oxidant for selenium, thus allowing the use catalytic amounts of SeO_2. Yields for this reaction are, however, variable a depend greatly on the structure of the alkene being oxidised (for exampl allylic oxidation of small–ring cyclic alkenes is not always an efficie process).

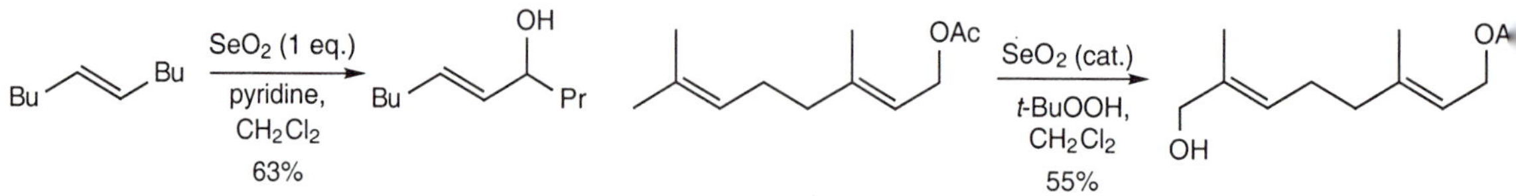

Bu–CH=CH–CH₂–Bu → SeO_2 (1 eq.), pyridine, CH_2Cl_2, 63% → Bu–CH=CH–CH(OH)–Pr

OAc substrate → SeO_2 (cat.), *t*-BuOOH, CH_2Cl_2, 55% → OH product

Fig. 4.12

The mechanism by which this reagent accomplishes oxidation has be studied in detail and proceeds *via* an initial ene-reaction between SeO_2 and th alkene. The resulting alkyl seleninic acid then rearranges *via* a [2, sigmatropic rearrangement and subsequent cleavage of the Se–O bond eith *in situ* or on work-up produces an allylic alcohol, Fig. 4.13.

Reactions with alkenes deuterated at the allylic positions show a primary kinetic isotope effect, which suggests that the initial ene reaction is rate determining.

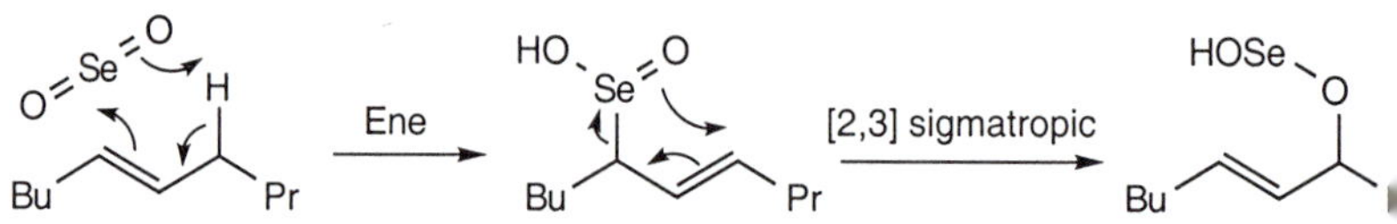

O=Se=O + Bu–CH=CH–CH₂–Pr → Ene → alkyl seleninic acid → [2,3] sigmatropic → Bu–CH=CH–CH(OSeOH)–

alkyl seleninic acid

Fig. 4.13

If you have trouble seeing the NBS–promoted bromination as an oxidation, consider that the bromide product in Fig. 4.14 can be transformed into an alcohol in aqueous solvent, *via* an S_N1 process. Clearly, the allylic carbon has been oxidised from level 0 to level 1 by this sequence.

N-Bromosuccinimide (NBS) is another reagent that is effective at oxidisi allylic positions, this time through a radical mechanism Fig. 4.14. NBS promoted bromination works best in a non-polar solvent, such as CCl_4, a with a small amount of radical initiator, such as AIBN.

The reaction proceeds *via* a radical mechanism which involves molecul bromine, formed *in situ*.

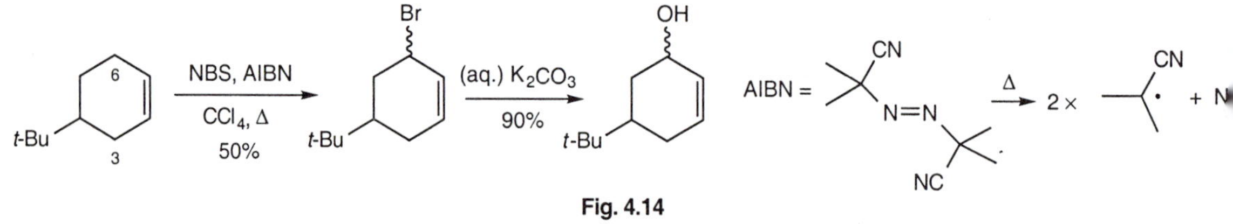

4-*t*-Bu-cyclohexene → NBS, AIBN, CCl_4, Δ, 50% → 3-bromo-5-*t*-Bu-cyclohexene → (aq.) K_2CO_3, 90% → 5-*t*-Bu-cyclohex-2-enol

AIBN = $(CH_3)_2C(CN)–N=N–C(CN)(CH_3)_2$ → Δ → 2 × $(CH_3)_2\dot{C}(CN)$ + N_2

Fig. 4.14

Under the reaction conditions, the initiator breaks down to form sma amounts of radicals in solution Fig. 4.14. In the first cycle, these radica abstract a hydrogen atom from the allylic position of the alkene, Fig. 4.1 The resulting allylic radical then reacts with bromine to form an allyl omide plus a Br radical. The reaction is then propagated by further drogen atom abstraction by the bromine radical, forming HBr in the ocess. The scheme is completed nicely by the observation that HBr reacts th NBS to form Br_2; so NBS acts as a slow releasing agent that generates a w concentration of Br_2 at any given time.

In fact, these low concentrations of Br_2 are crucial in preventing addition of $_2$ to the alkene and allow allylic substitution to compete successfully. The n-polar solvent also disfavours bromination of the alkene by Br_2 *via* a omonium ion (as this polar reaction gives rise to charged intermediates and, t surprisingly, these are disfavoured in CCl_4).

The regiochemistry of the bromination shown in Fig. 4.15 is determined by preferential hydrogen atom abstraction at the C-6 position over the more hindered C-3 position.

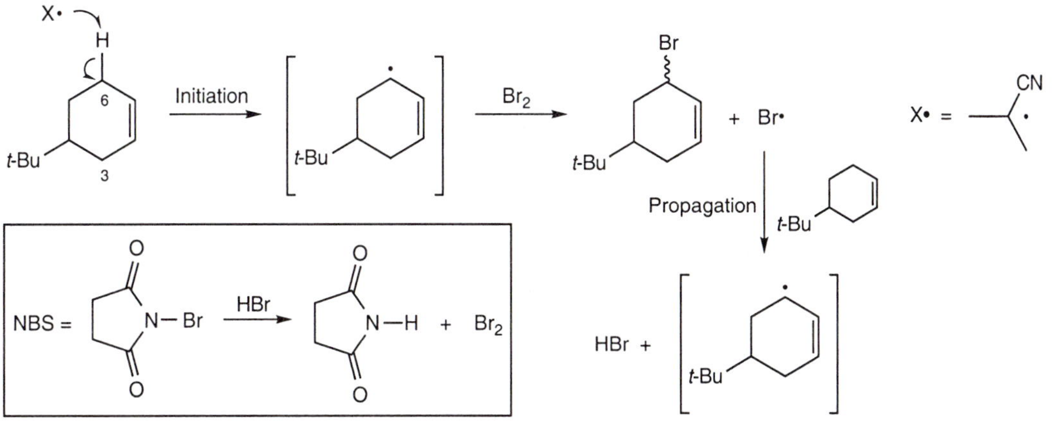

Fig. 4.15

xidation at benzylic positions

rbon–hydrogen bonds that are adjacent to an aromatic system (*eg*. benzylic) also activated and can be removed in an oxidative process. Chromic acid $_2CrO_4$) and sodium dichromate ($Na_2Cr_2O_7$) are reagents capable of oxidising omatic methyl groups all the way to carboxylic acids, Fig 4.16. Obviously presence of other oxidisable functionality (alcohols, alkenes, *etc.*) should avoided if this reaction is to be successful. Oxidation of the pyridine own in Fig. 4.16) to the corresponding *N*–oxide is probably slow because amine is protonated under the acidic conditions.

The mechanism by which these oxidations proceed is undoubtedly complex. Initial oxidation may take place by abstraction of a hydrogen atom or a hydride ion from the benzylic carbon.

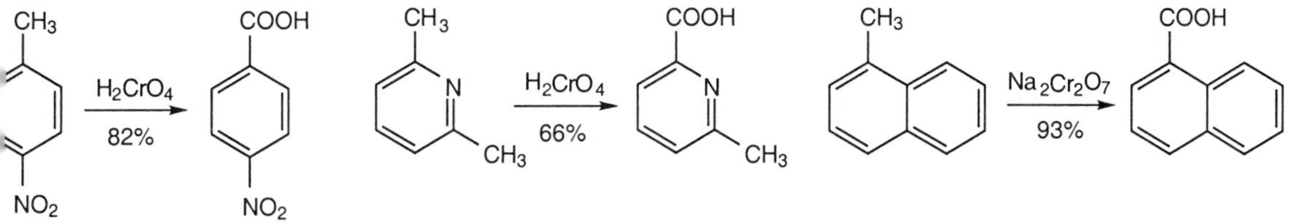

Fig 4.16

3 Oxidation adjacent to a carbonyl group

e transformation of an ester or a ketone into an α–hydroxy–carbonyl mpound is a useful one to learn, not least because the hydroxy–carbonyl otif is found in many natural products. How might one introduce a hydroxyl group adjacent to a carbonyl group? You will discover that all the oxidations that we shall introduce make use of an enol or an enolate ani as an intermediate (these are normally formed *in situ* rather than prepared in separate step). With this in mind we could just as easily have put this secti in Chapter 3, as enols and enolates are simply electron-rich alkenes and so shouldn't be surprising to find that they are oxidised with relative ease.

Oxidation of enolates

Obviously, one can make an enolate by reacting a carbonyl compound wi strong base; what we then need is an *electrophilic* source of oxygen for t enolate to react with. The simplest reagent to fulfil this role is gaseo oxygen, Fig. 4.17, and the result of such a combination is formation of hydroxy ketones in good yields.

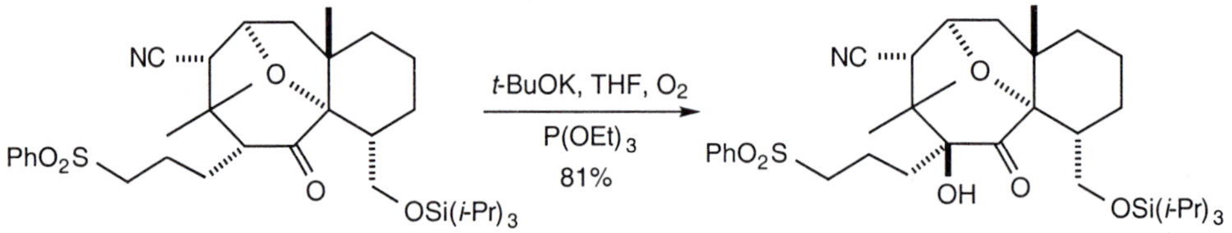

t-BuOK, THF, O_2
$P(OEt)_3$
81%

Fig. 4.17

Reactive potassium enolates derived from ketones work best with th procedure, which is presumed to involve radicals, Fig. 4.18. Initial oxidati of the enolate by molecular oxygen generates an α-keto radical which c combine with triplet oxygen in another radical reaction (remember th ground–state oxygen is a triplet diradical). The resulting peroxy radical th propagates the chain by abstracting an electron from another molecule enolate. A peroxide anion is formed and this can then be protonated. Un normal circumstances the peroxide product is cleaved during work-up reduction of the weak O–O bond with zinc or triethylphosphite.

The peroxide product may eliminate water if a proton is present at the α-position. This side-reaction is often observed and leads to overoxidation. Consequently, molecular oxygen is at its most effective when oxidising tertiary centres adjacent to ketones as the peroxide cannot eliminate.

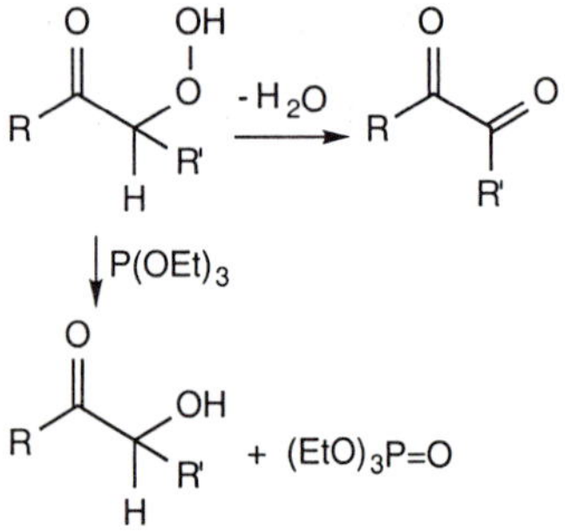

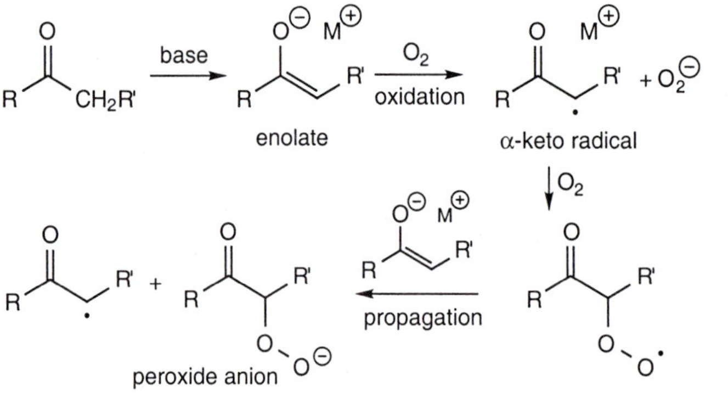

Fig. 4.18

N-Sulfonyloxaziridines as an electrophilic source of oxygen

N-Sulfonyloxaziridines are stable compounds that react (as electrophiles) w enolate nucleophiles, thus forming α-hydroxy-carbonyl compounds in o step. The electrophilic nature of *N*-sulfonyloxaziridines is partly due to arge electron-withdrawing nature of the SO_2 group and nucleophiles open the trained ring by reaction at the unhindered oxygen atom, Fig. 4.19. Hydroxyl roups may be introduced adjacent to ketones, esters and lactones by first eprotonating them with a strong base, KHMDS = $(Me_3Si)_2N$–K, and then sing this reagent as an electrophile.

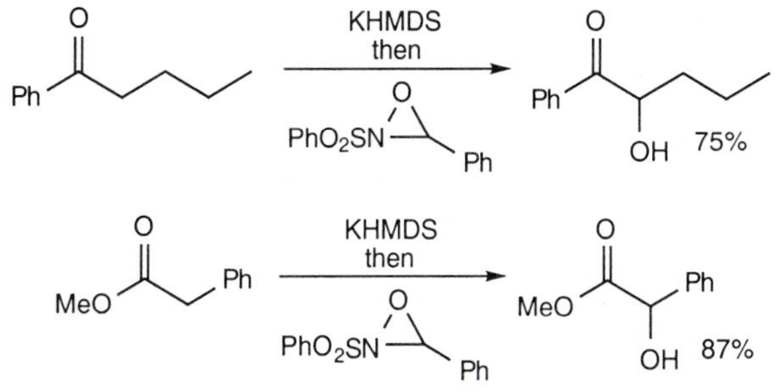

The overall displacement can be viewed thus; viewing oxygen as an electrophile is rather an unusual process.

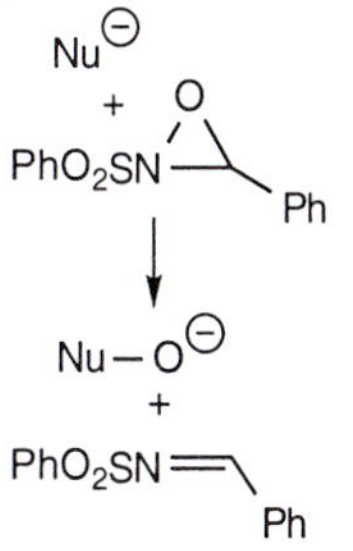

Fig. 4.19

.4 Oxidation adjacent to nitrogen

he oxidation (by dehydrogenation) of amine to an imine is not a particularly asy transformation to achieve in organic chemistry. One particularly dmirable oxidising agent for a related oxidation is the di-*t*-butyl–substituted uinone **8**, Fig. 4.20. This compound is effective in transforming primary mines into imines which are then hydrolysed *in situ* to the corresponding etones (unfortunately one cannot transform RCH_2NH_2 into an aldehyde *via* is chemistry).

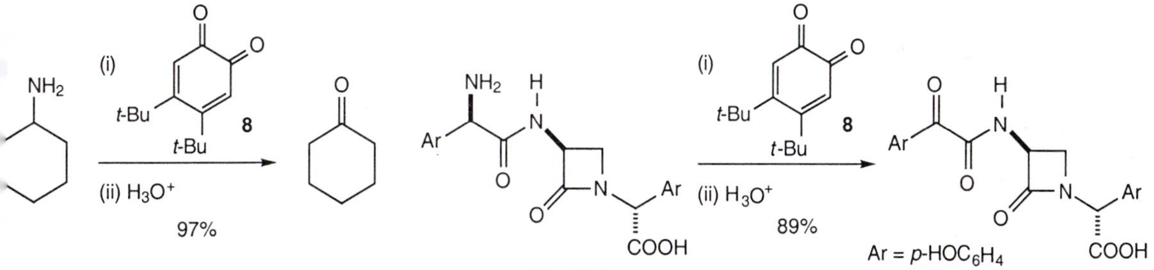

Fig. 4.20

This oxidation proceeds *via* initial condensation of the amine and quinone rming an imine, Fig. 4.21.

The two *t*-butyl groups in **8** are there to prevent a sidereaction, namely conjugate attack of the amine onto the quinone.

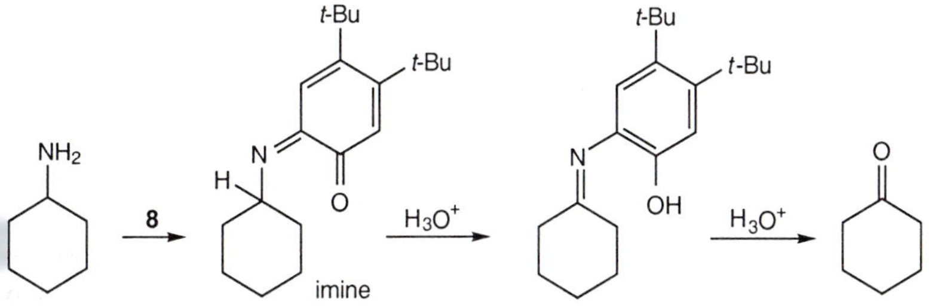

Fig. 4.21

The second step, which uses aqueous acid, begins with a tautomerisatio to the thermodynamically more stable isomer (presumably the driving force i formation of an aromatic ring). Subsequent hydrolysis of the resulting imin furnishes a ketone; just as the amine has been oxidised, the quinone has bee reduced to an aminophenol.

The chemistry described above is based on the way that nature oxidise amino acids to keto-acids using pyridoxal phosphate **9**. The overall equatio for conversion of an amino acid (alanine is shown as an example) to a ket acid (pyruvate in this case) is illustrated below and is called transaminatio with glutarate/ketoglutarate as a reaction partner, Fig. 4.22.

Obviously, amino acids are ingested whenever one eats protein; although the amino acid monomers from protein are recycled, excess material is degraded. Other biosynthetic pathways that require the oxidation of an amino group to a carbonyl may use similar chemistry.

The enzymes that catalyse the transamination pathway are, no surprisingly, called transamidases and they use pyridoxal phosphate as a essential co-factor. The co-factor is tightly bound by the enzyme in the activ site where it can participate in the ensuing reaction. In essence, the primar amine group of an amino acid (say alanine) condenses with the aldehyd functional group of **9**, Fig. 4.23 (compare with Fig. 4.21). Rearrangemen ensues and the imine **10** that is so formed can then be hydrolysed to a keto acid (namely pyruvate).

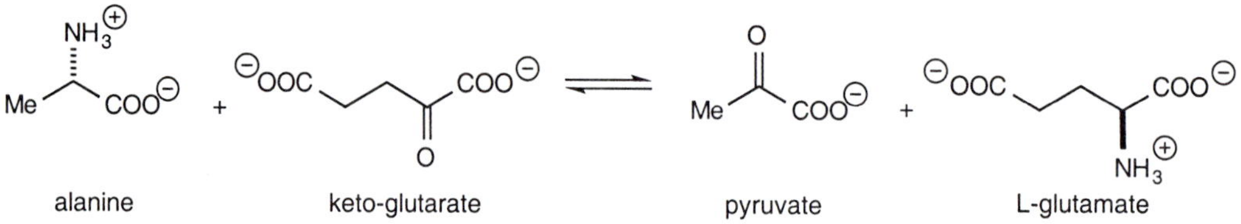

alanine + keto-glutarate ⇌ pyruvate + L-glutamate

Fig. 4.22 Transamination

A hydrogen bond between the C-3 hydroxyl group and the nitrogen of the imine is thought to ensure co-planarity of the delocalised π-system so ensuring efficient proton transfer.

Clearly, if an amino acid is oxidised then pyridoxal must be reduced (t pyridoxamine in this case). In order that pyridoxal may be used again pyridoxamine is subsequently re-oxidised *in vivo* in another process which i essentially the reverse of Fig. 4.23 but which involves the keto-glutarate t glutamate couple.

9 + alanine ⇌ ⇌ **10** ⇌ (H_2O) pyridoxamine + pyruvate

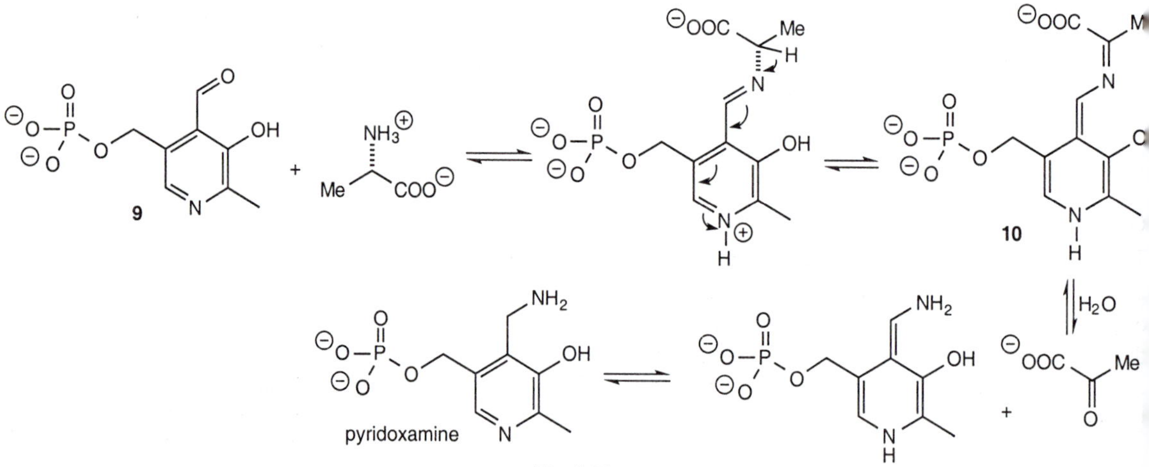

Fig. 4.23

4.5 Oxidation of phenols: formation of quinones

Oxidation of a phenol can give rise to a quinone *via* removal of the two hydrogen atoms attached to oxygen: this is clearly a dehydrogenation (which is one of our definitions of oxidation). The hydrogen atoms that are removed are benzylic and, therefore, activated by the aromatic ring.

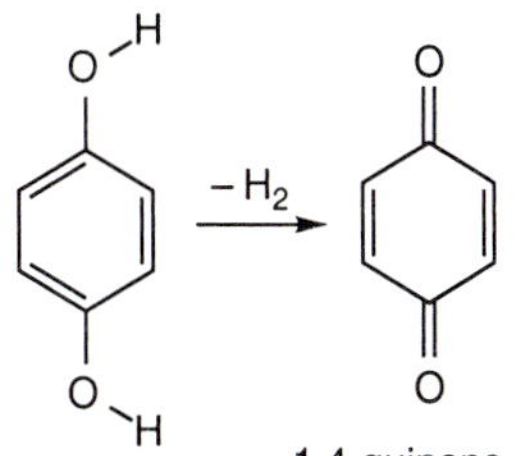

Quinones are important compounds that are used in dyestuffs and are also found in many natural products. There are three main methods for synthesising quinones that will be considered here:

$- H_2$

1,4-quinone

Oxidation of hydroquinones

The oxidation of 1,2- or 1,4-dihydroxybenzenes (these are also known as hydroquinones) to quinones is usually an easy oxidation to accomplish. In keeping with this observation there are many oxidants that can be used to prepare quinones. Two of the most useful are silver carbonate (silver 1+ is reduced to silver metal) and activated manganese dioxide, Fig. 4.24.

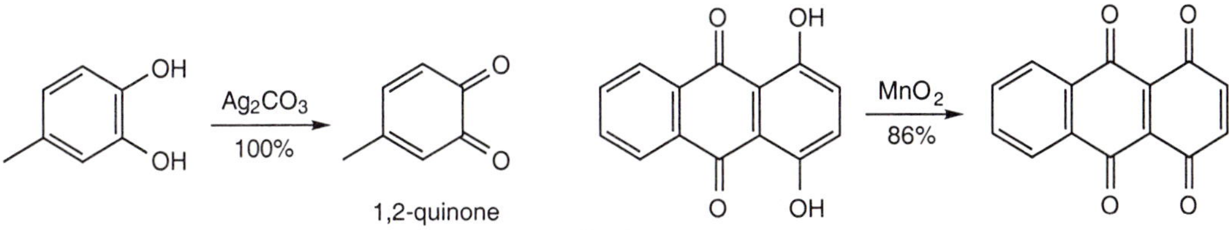

Ag_2CO_3 100%

1,2-quinone

MnO_2 86%

Fig. 4.24

Generally, it is easier to oxidise hydroquinones that are substituted with electron-donating groups. Consequently, hydroquinones that are substituted with electron-withdrawing groups are more difficult to oxidise. Clearly, electron-deficient quinones are easily reduced to the corresponding hydroquinone and so can be used as oxidising agents themselves. For example, (excess) 2,3-dichloro-5,6-dicyanoquinone, DDQ, Fig. 4.25, is capable of dehydrogenating the diol shown (– $3H_2$) to a quinone.

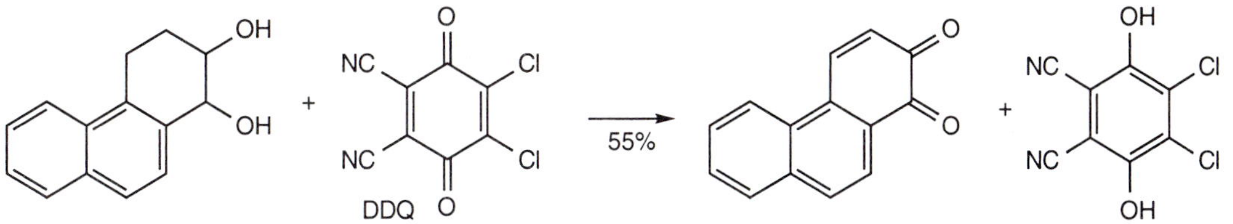

DDQ

55%

Fig. 4.25

Oxidative demethylation

The transformation of 1,4-dimethoxybenzene into a quinone could rightly be considered in Chapter 5, as it involves an oxidative cleavage of a carbon–oxygen sigma bond; however, I have included the reaction at this juncture so that you can appreciate it as another method of forming quinones.

So, reaction of **11** with ceric ammonium nitrate [$(NH_4)_2Ce(NO_3)_6$ or CAN] is a useful method of forming the corresponding quinone, Fig. 4.26.

CAN is a good one-electron oxidant. During this process the cerium 4^+ ion gains an electron and is so reduced to cerium 3^+.

The use of ^{18}O labelled water in this oxidation (Fig. 4.26) confirmed that the oxygen atoms on the quinone come from the water solvent and do not originate from the starting material.

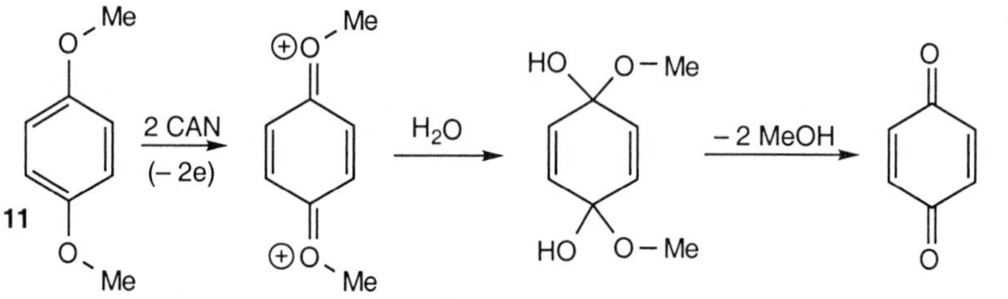

Fig. 4.26

The mechanism of reaction involves the loss of two electrons from the aromatic ring (two molecules of CAN are reduced in this scheme) and nucleophilic attack onto the cation by two moles of water. Loss of methanol (× 2) completes the sequence and produces the quinone.

Oxidation with introduction of oxygen

Oxidation of phenols (with Fremy's salt) that are blocked at the C-4 position leads to 1,2-quinones.

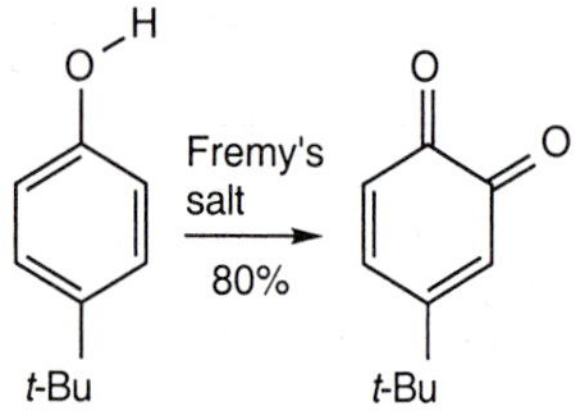

Phenols with one hydroxy group may also be oxidised to quinones by the introduction of another oxygen atom. The best reagent for achieving this transformation is Fremy's salt (**12**, Fig. 4.27). This stable free radical reacts with phenol to produce the corresponding 1,4-quinone (rather than the 1,2-quinone) in good yield. In terms of mechanism, the first step is presumed to be hydrogen atom abstraction by **12** to give a benzylic (and therefore stabilised) radical; this may then combine with another mole of Fremy's salt to give **13**. This recombination pathway occurs more readily at C-4 than at the more hindered C-2 position, Fig. 4.27. Finally, elimination of the electron-deficient nitrogen occurs so furnishing a quinone.

Reaction with ^{18}O labelled **12** confirmed that the oxygen introduced at C-4 did indeed come from the salt, as we would expect from the mechanism shown.

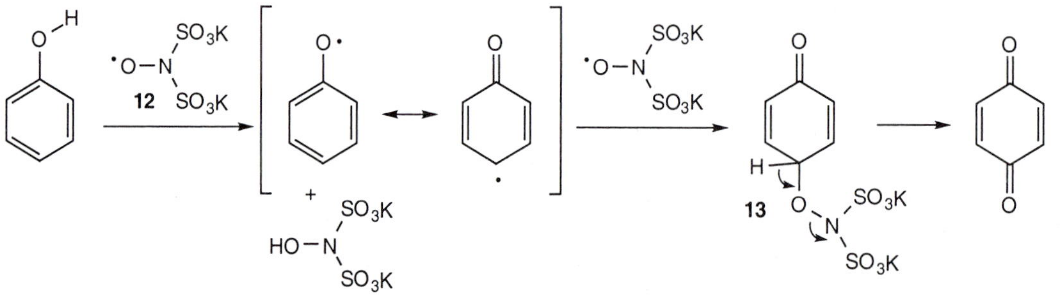

Fig. 4.27

Compound **12** is purple and this colour fades as the reaction reaches completion.

5. Oxidative cleavage reactions

In this chapter we shall concentrate on reactions which lead to cleavage of carbon–carbon single (and multiple) bonds and the introduction of new bonds between carbon and an electronegative element, such as oxygen. For completeness we will also consider a similar process which leads to the cleavage of bonds between carbon and other electropositive elements such as boron or silicon.

5.1 Oxidative cleavage of carbon–carbon double bonds

Oxidative cleavage of alkenes using ozone

Clipping an alkene to furnish two carbonyl compounds is most readily carried out with ozone (O_3): the four substituents that were attached to the alkene end up as substituents on the two new carbonyl containing compounds. This reagent is extremely specific for alkenes and so ozonolysis can be performed on multi-functional compounds without fear of oxidising (for example) alcohols, esters or aromatic rings, Fig. 5.1.

Cleavage of an alkene takes place in two stages: first, the alkene (CH_2Cl_2 or MeOH are good solvents) is cooled to low temperature and ozone is passed through the solution. This gives rise to an ozonide (*vide infra*) which is then decomposed in a separate step using dimethylsulfide.

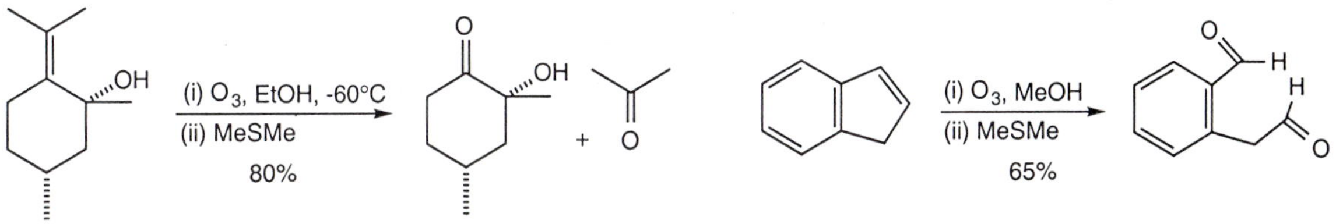

(i) O_3, EtOH, -60°C
(ii) MeSMe
80%

(i) O_3, MeOH
(ii) MeSMe
65%

Fig. 5.1

Ozone is an electrophilic reagent and so reacts fastest with electron-rich alkenes. However, the differences in rate for ozonolysis of alkenes with varying numbers of alkyl groups are small and it is not usually possible to distinguish between them. What is possible, however, is the oxidation of an alkyl–substituted alkene in the presence of a carbonyl-substituted (and therefore very electron–deficient) alkene, Fig. 5.2.

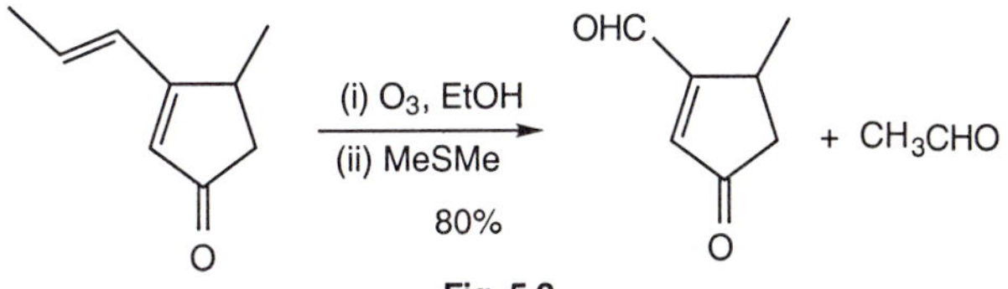

(i) O_3, EtOH
(ii) MeSMe
80%
+ CH_3CHO

Fig. 5.2

The mechanism by which alkenes react with ozone has been studied in some detail, Fig 5.3. In the first stage, ozone participates in a cycloaddition reaction with the alkene. This pericyclic reaction involves four π electrons from ozone and two from the alkene and is therefore allowed under the Woodward–Hoffmann rules (as long as it is suprafacial). And, as ozone is a [1,3] dipole, the reaction may be classified as a [1,3] dipolar cycloaddition. The resulting complex (primary ozonide) is unstable and collapses *via* another pericyclic reaction; this time it is a reverse [1,3] cycloaddition which cuts the ozonide in another way. A carbonyl compound and a carbonyl ylid are formed by this path and these may combine with either the solvent, path A, or with each other, path B.

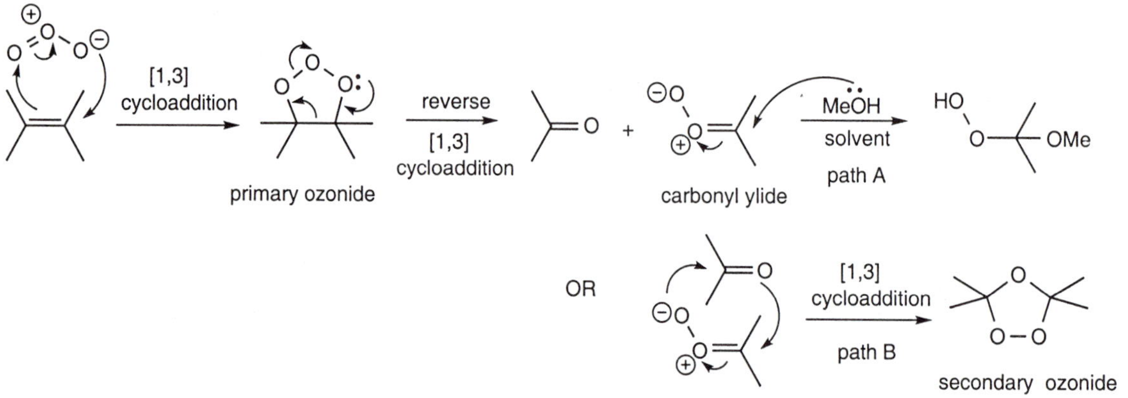

Fig. 5.3

The peroxides and secondary ozonide formed by this mechanism are almost never isolated but instead are decomposed *in situ* by the addition of dimethyl sulfide (Me_2S) or trimethyl phosphite ($[MeO]_3P$), Fig. 5.4. In this process either the sulfur or the phosphorus additive is oxidised and the ozonide or peroxide reduced to the corresponding carbonyl compound.

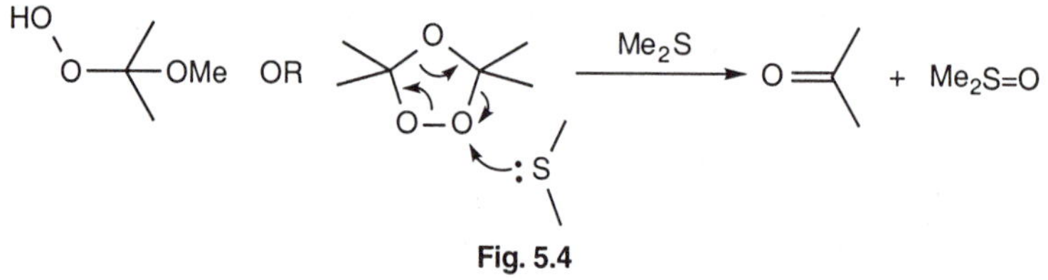

Fig. 5.4

It is possible to decompose secondary ozonides in a variety of ways. For example, reaction with sodium borohydride, instead of Me_2S, reduces the ozonide to give alcohols, Fig. 5.5. Conversely, one can also cleave certain ozonides by oxidising them with hydrogen peroxide to form carboxylic acids.

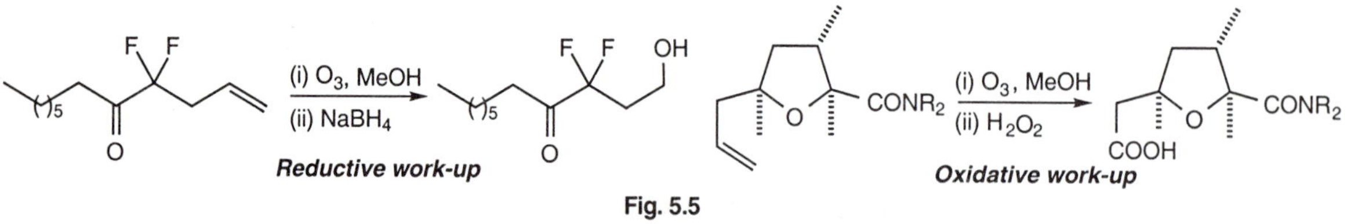

Fig. 5.5

Oxidative cleavage of aromatic rings

We have already come across ruthenium tetroxide as a powerful oxidant capable of transforming alkynes into 1,2-diketones. This oxidant will also cleave alkenes in a manner reminiscent of ozone (remember that RuO_4 is made *in situ* by the oxidation of $RuCl_3$ or RuO_2 with $NaIO_4$). Another useful oxidation that ruthenium tetroxide promotes is the cleavage of aromatic rings to yield carboxylic acids: both carbocyclic and heterocyclic aromatic rings can be cleaved, Fig. 5.6. Clearly, this cleavage reaction cannot be performed in the presence of other easily oxidisable groups such as alkenes or alcohols.

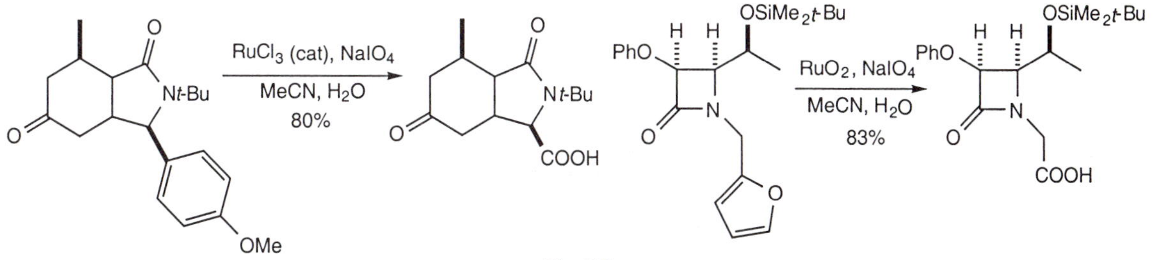

RuCl3 (cat), NaIO4, MeCN, H2O, 80%

RuO2, NaIO4, MeCN, H2O, 83%

Fig. 5.6

By using this reaction one can think of benzene or furan rings as masked forms of carboxylic acids!

In terms of a mechanism, one can imagine ruthenium tetroxide adding to each of the π-bonds within the arene ring in a manner akin to osmium tetroxide (presumably electron-rich rings react fastest). The resulting intermediates may then be cleaved either by the ruthenium oxidant or sodium periodate, *vide infra*.

5.2 Oxidative cleavage of carbon–carbon sigma bonds

Cleavage of 1,2-diols using periodate

1,2-Diols can be oxidatively cleaved to carbonyl compounds by sodium periodate in methanol (sometimes periodic acid H_5IO_6 is used instead). This reaction is mild, high yielding and extremely specific for diols that are adjacent to each other, so making it useful in organic synthesis, Fig. 5.7. It does not matter if the diol unit is part of a ring or not: cleavage occurs just the same. Imagine cleaving a polyhydroxylated substrate (such as the carbohydrate glucose) with excess periodate, Fig. 5.7. Each 1,2-diol will be cleaved and then each α-hydroxy carbonyl group will be cleaved again— this method is, therefore, one of the best ways of selectively degrading sugars.

You have seen $NaIO_4$ before, as a reagent for oxidising sulfides to sulfoxides and selenides to selenoxides. Unfortunately, these easily oxidisable groups will interfere with the diol cleavage reaction.

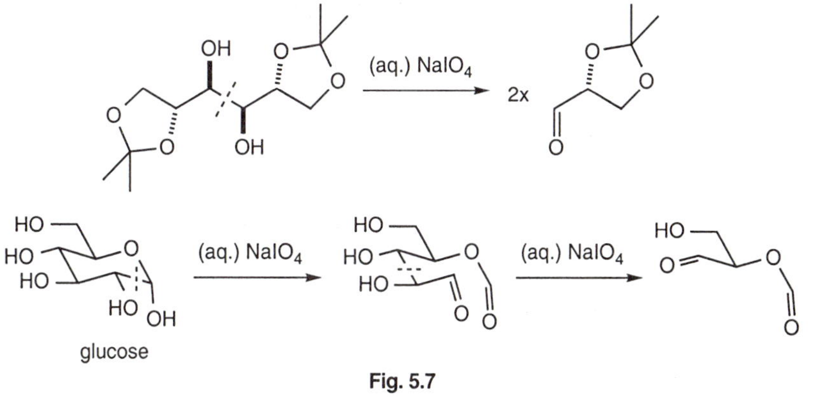

(aq.) NaIO4 → 2x

glucose (aq.) NaIO4 → (aq.) NaIO4 →

Fig. 5.7

Other functionality related to the 1,2-diol unit may also be oxidatively cleaved under these conditions. So, α-hydroxy carbonyl compounds may also be cleaved to yield aldehydes and carboxylic acids, Fig. 5.8.

It is easiest to consider cleavage proceding *via* the hydrated form of the ketone: this is simply a 1,2–diol, formed *in situ.*

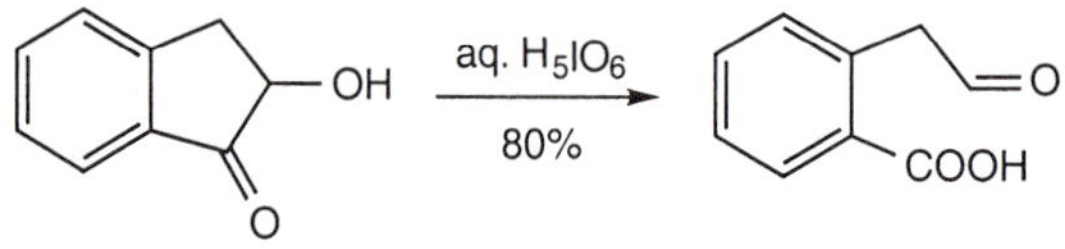

aq. H_5IO_6 80%

Fig. 5.8

The mechanism of such cleavage reactions involves formation of a chelate between the periodate and the diol; this may then collapse as shown, forming two carbonyl compounds and reducing iodine in the process, Fig. 5.9. Hence, diols that cannot form a chelate are inert to this method of oxidation (see **1**).

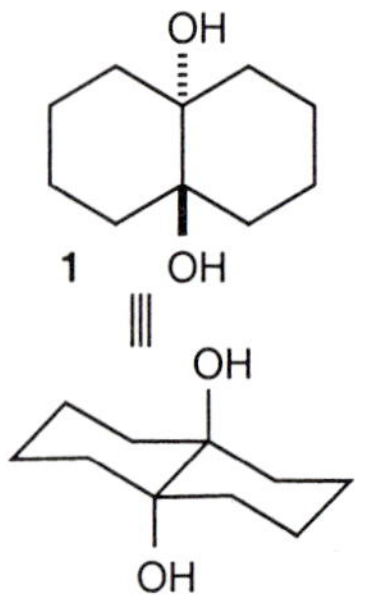

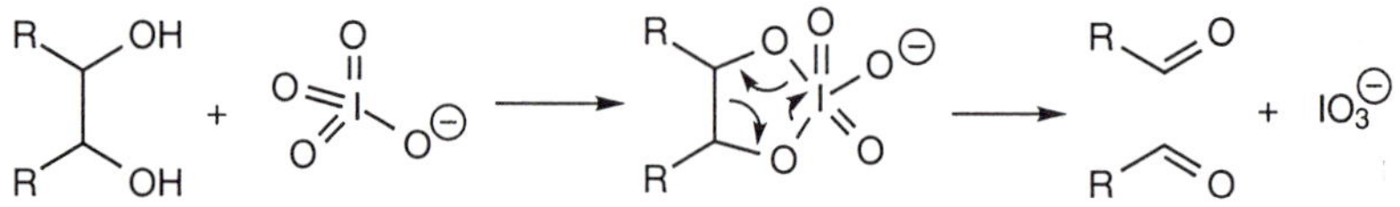

Fig. 5.9

Lead tetraacetate is another reagent that will cleave 1,2-diols and is often used instead of periodate. Although the mechanism by which it acts is probably similar, it is less important that the diol forms a chelate and so compound **1** *would* be cleaved slowly into a diketone with this reagent.

Formation and cleavage of 1,2-diols in one step

You may remember that *cis*-1,2-diols are easy to make from alkenes using osmium tetroxide (see Chapter 3). If we combine osmium tetroxide and sodium periodate in one reaction vessel then an interesting reaction ensues. First, osmium tetroxide transforms the alkene into a 1,2-diol and, second, periodate cleaves the 1,2-diol into a dicarbonyl compound (as described above, Fig. 5.10). Even better is the fact that catalytic osmium tetroxide can be used in this sequence because the periodate acts as a re-oxidant for the osmium, regenerating osmium tetroxide *in situ*. So, we can consider the mixture of osmium tetroxide and sodium periodate as an alternative to ozonolysis.

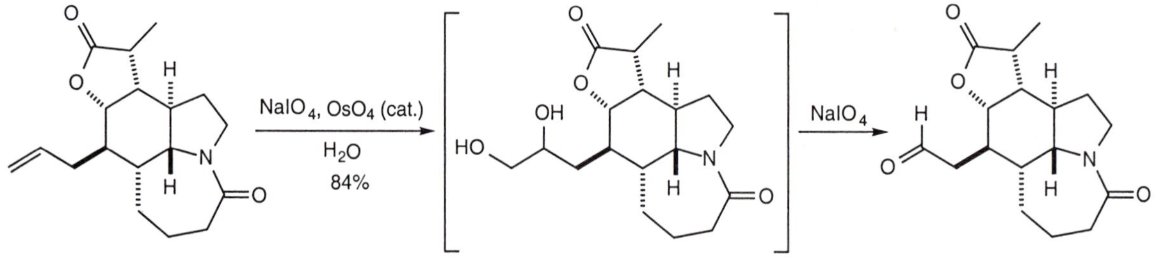

$NaIO_4$, OsO_4 (cat.) H_2O 84%; $NaIO_4$

Fig. 5.10

Cleavage of carbon–carbon single bonds adjacent to a carbonyl group (the Baeyer–Villiger reaction)

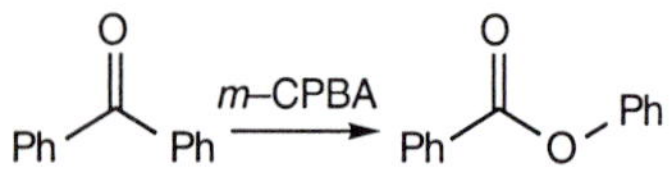

Reaction of an aldehyde or a ketone with a peracid results in an oxidative cleavage of one of the substituents attached to the carbonyl group; this is known as the Baeyer–Villiger reaction. *m*-CPBA or trifluoroperacetic acid (CF_3COOOH) are the best known reagents and the latter is particularly effective at oxidising unreactive substrates.

m–CPBA

the Baeyer–Villiger reaction

In order to understand how more complex systems behave, it is instructive to consider the mechanism of the oxidation. It is clear that the reason behind the specific cleavage of sigma bonds adjacent to carbonyls is related to the fact that peracids can act as nucleophiles towards the C=O group, Fig. 5.11. As one would expect, formation of the adduct **2** is preceded by initial protonation of the carbonyl group (so increasing its electrophilicity). Now comes the step that actually cleaves the carbon–carbon sigma bond; this involves the lone pair on the hydroxyl group of **2** assisting a carbon atom to migrate towards the peroxide linkage. In so doing, the weak O–O bond is cleaved and a stabilised oxy-anion formed as a leaving group. In fact it is accepted that the migration step is rate determining and this explains why CF_3COOOH, which leads to the expulsion of a particularly stable leaving group CF_3COO^-, is so good at promoting the reaction.

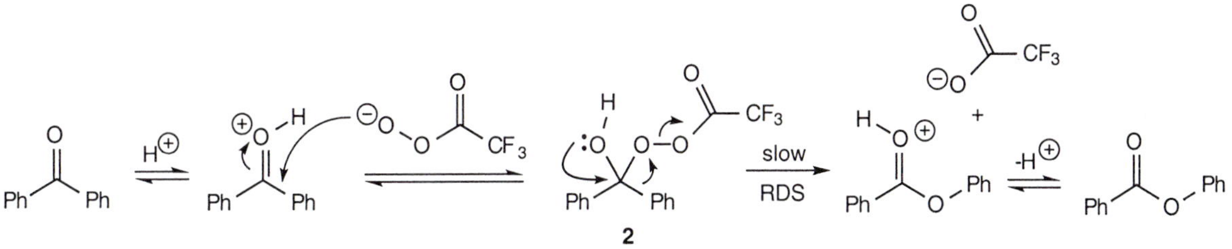

Fig. 5.11

What happens when we oxidise unsymmetrical ketones using the Baeyer–Villiger reaction? Clearly two products are possible, depending upon which carbon atom migrates. Consider oxidation of *t*-butylethyl ketone **3**, Fig. 5.12. Migration of the ethyl group would lead to compound **4**, and migration of the *tert*-butyl group to **5**. Fortunately, the migration process is a rather predictable one and an order of migrating group ability has been constructed. This order places *more–substituted* carbon as the *best* migrating group and CH_3 as the worst migrating group (tertiary carbon > secondary ≈ aryl > primary > methyl). This means that **5** would be the expected product from the reaction shown in Fig. 5.12, and this is indeed the case.

This mechanism is supported by the oxidation of ^{18}O labelled benzophenone. The product ester incorporates the label exclusively in the doubly bonded oxygen.

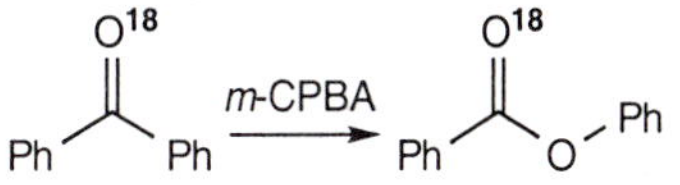

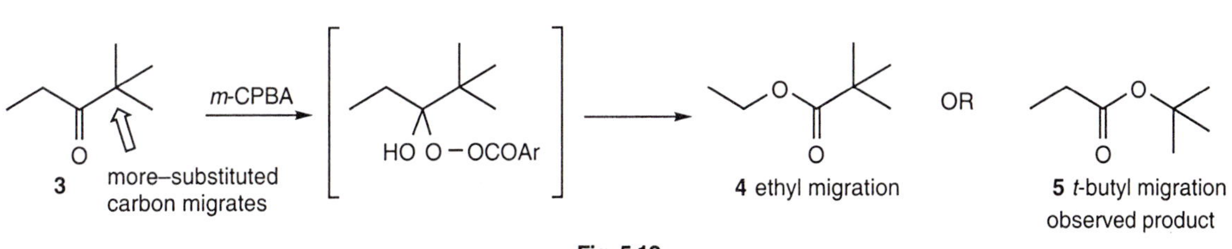

Fig. 5.12

Several examples of this type of selectivity are shown in Fig. 5.13 where you can see a phenyl group competing for migration (and winning) over that of a methyl and an ethyl group (Ph > Me, Ph > Et). However, as the phenyl group is approximately equal in migrating ability to an *iso*-propyl group, mixtures are formed (Ph ~ *i*–Pr) and, indeed, the *tert*-butyl group is even better at migrating than a phenyl (*t*–Bu > Ph). Moreover, aldehydes normally react *via* hydrogen migration, thus forming a carboxylic acid in the process.

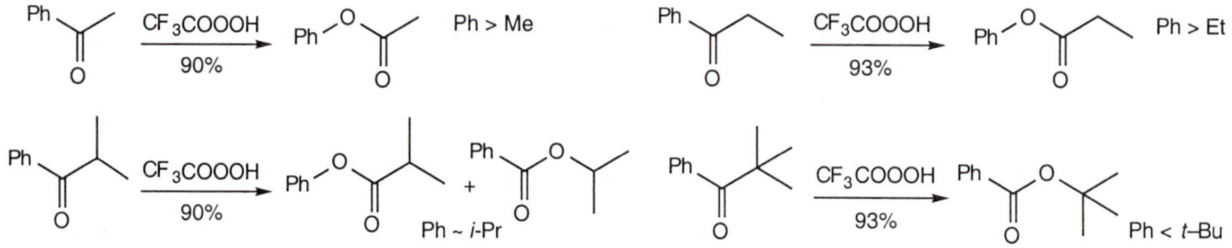

Fig. 5.13

The reason behind the order of migratory aptitude is unclear. It appears that more substituted carbons migrate (to an electron–deficient oxygen) because they can better sustain a partial positive charge (remember that tertiary carbocations are more stable than secondary, etc.). Moreover, it is likely that the regiochemistry of rearrangement is influenced by the greater relief of steric strain that accompanies migration of a bulky group and also by stereoelectronic effects.

What happens if the carbon that migrates is a stereogenic centre? Again, this aspect of the Baeyer–Villiger reaction has been studied in some detail and carbon migrates with retention of configuration, Fig. 5.14. Although stabilisation of positive charge is important, simplistically, carbon migrates with its pair of electrons. So, for example, in the oxidation of **6**, the carbon that migrates keeps its substituents in the same configuration.

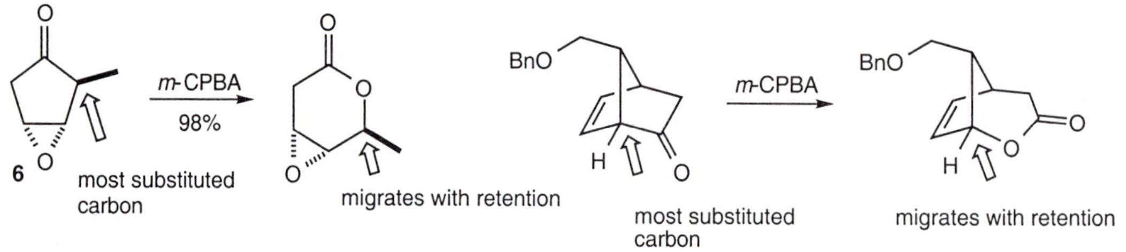

Fig. 5.14

5.3 Oxidative cleavage of carbon–boron and carbon–silicon bonds

Bonds between boron (and silicon) and carbon may also be cleaved oxidatively in a process that is rather similar to the Baeyer–Villiger reaction. Remember that both boron and silicon are more electropositive elements than carbon and so we can consider the transformation of C–B (and C–Si) into C–O as an oxidation.

The most common way to make C–B bonds is to add the H–B bond across an alkene (hydroboration with BH_3). Reaction of the products from a hydroboration reaction with alkaline hydrogen peroxide causes the C–B bond to rupture and migrate onto oxygen, Fig. 5.15.

The cleavage reaction is initiated by attack of $^{-}$OOH on boron (the nucleophile can add to the empty p-orbital on boron). The resulting complex is negatively charged and, in order for the boron to regain neutrality, a carbon attached to the boron undergoes migration onto the peroxide linkage, displacing $^{-}$OH in the process. Hydrolysis of the resulting boron–oxygen bond occurs readily and the product is an alcohol.

It is interesting to note that the sequence of hydroboration followed by hydrogen peroxide treatment does not constitute either an oxidation or a reduction (both starting material and product are in oxidation level 1). We can consider this two–step process as a reduction (addition of B–H, to be covered later on) followed by an oxidation, which gives no overall change in oxidation level.

The carbon–boron bond can be oxidised further than an alcohol and directly into a carbonyl group by reaction with a stronger oxidant than H_2O_2, namely chromic acid (H_3CrO_4), Fig. 5.15.

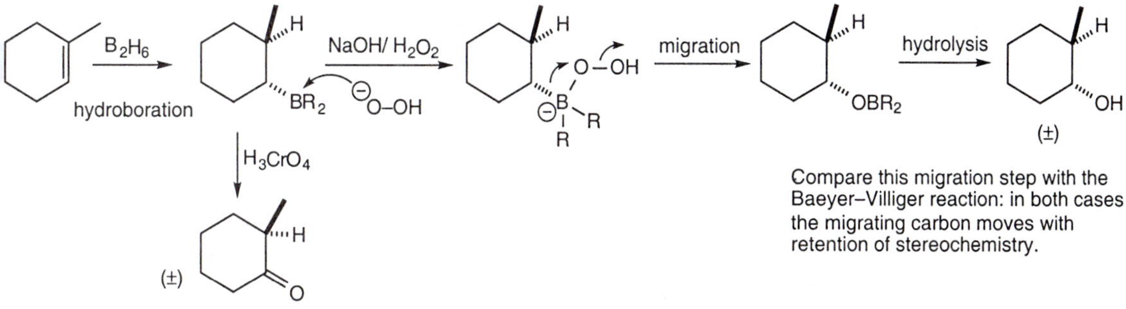

Compare this migration step with the Baeyer–Villiger reaction: in both cases the migrating carbon moves with retention of stereochemistry.

Fig. 5.15

Carbon–silicon bonds may also be cleaved oxidatively in a manner similar to boron. In order for the reaction to work, it is necessary to have an electronegative substituent (such as oxygen or fluorine) on the silicon to increase its electrophilicity. So, compound **7** cannot be cleaved to the alcohol until the phenyl group on silicon has been replaced by a fluorine by protodesilylation with HBF_4, Fig. 5.16. A plausible mechanism involves nucleophilic attack of the anion of *m*-CPBA onto the electrophilic silicon atom. This process generates a negatively charged ATE complex of silicon. In a reaction reminiscent of that described for the oxidation of C–B bonds, the carbon attached to silicon migrates onto the peroxide linkage, displacing a carboxylate leaving group and restoring electrical neutrality to silicon. Hydrolysis of the oxygen–silicon bond occurs readily to yield the alcohol product.

In this chapter we have seen several examples of carbon migrating with retention of stereochemistry. Not surprisingly, this is also the case with migration of carbon from silicon onto oxygen.

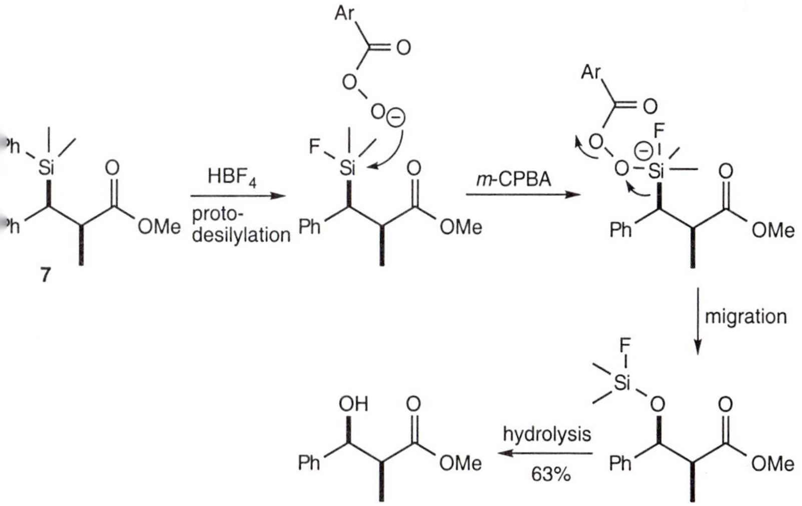

Fig. 5.16

5.4 Oxidative cleavage of carbon–halogen bonds: the Kornblum oxidation

A useful method of oxidising carbon–halogen bonds involves reaction of alkyl halides with dimethylsulfoxide (DMSO) in the presence of a base (usually $NaHCO_3$ or Et_3N): this is known as the Kornblum oxidation, Fig 5.17. In the reaction shown, benzyl bromide (carbon in oxidation level 1) is oxidised to benzaldehyde (level 2). The reaction initiates with nucleophilic attack by the sulfoxide oxygen onto the carbon–halogen bond in an S_N2 reaction. You have seen the product of this displacement **8** before! It is a sulfonium ion, last seen as an intermediate in the Swern oxidation (Chapter 4). Not surprisingly, **8** collapses to give a carbonyl compound and dimethyl sulfide— deprotonation is effected by the base which is added to the system.

It is instructive to consider the fate of the DMSO used in the reaction. Fig. 5.17 shows that it is reduced to dimethylsulfide as the alkylhalide is oxidised. There is no reason why we should not think of this reaction as a method of reducing sulfoxides to sulfides!

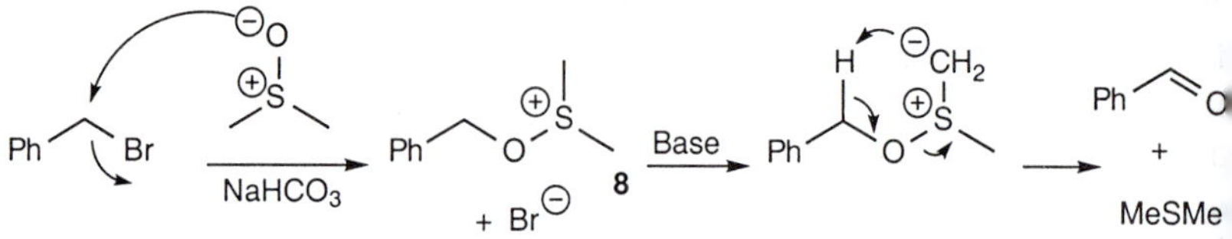

Ph Br NaHCO3 Ph O 8 + Br Base Ph O H CH2 Ph O + MeSMe

Fig. 5.17

If we are to expect the Kornblum reaction to work well then it is clear that the alkylhalide used must be able to undergo displacement easily consequently this oxidation is effective at transforming activated alkylhalides (i.e. carbon–halogen bonds adjacent to alkenes, arenes and carbonyl compounds) into aldehydes and ketones. A modification of the reaction reported by Kornblum, involves treatment of a primary, unactivated alkylhalide with silver tosylate (presumably to first make the corresponding tosylate), followed by DMSO and $NaHCO_3$ as before, Fig. 5.18.

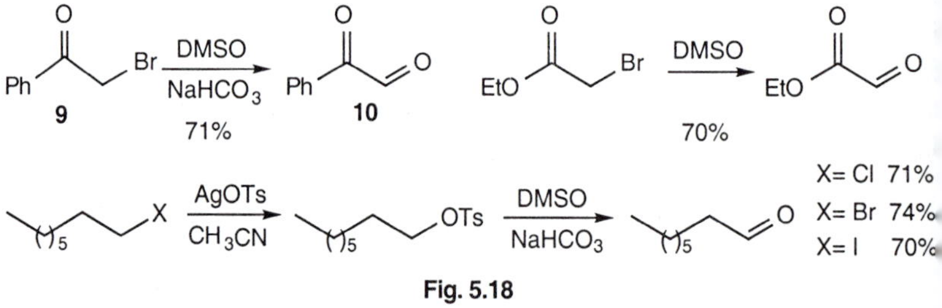

Ph Br 9 DMSO NaHCO3 71% Ph O 10; EtO Br DMSO 70% EtO O

X AgOTs CH3CN OTs DMSO NaHCO3 O; X= Cl 71%, X= Br 74%, X= I 70%

Fig. 5.18

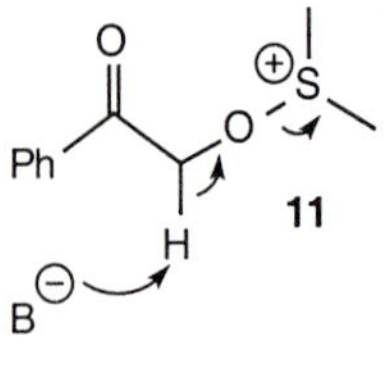

When an α-halo–ketone, such as **9**, is oxidised under the Kornblum conditions then the mechanism is subtly different from that described in Fig 5.17; the sulphonium ion **11** is not deprotonated to form an ylid but, instead loses one of the particularly acidic protons α to the carbonyl group to go directly to the product **10**.

Ph O H B 11

6. Reduction of heteroatoms attached to carbon

n Chapter 2 we considered oxidation of the heteroatoms N, P, S and Se and aw that these nucleophilic elements were easily oxidised by electrophilic xidising agents. For completeness, we will now consider the reverse eaction, i.e. reduction of heteroatom-based functional groups. In nearly all uch cases the reduction reaction manifests itself as a reduction in the number f oxygen atoms attached to a particular heteroatom. You will meet all anner of different reducing agents in the following chapters and I have ttempted to include one of each of the most important classes in Chapter 6. 'or example, you will come across reducing agents that deliver H_2 (either irectly or stepwise as H^- and H^+) to a compound. Other reducing agents will dd electrons to a substrate and there are yet more subtle reducing agents, with n affinity for oxygen, which will deoxygenate certain compounds.

.1 Reduction of nitrogen-containing functional groups

Reduction of nitro groups to amines

Heterogeneous catalysts are insoluble in the reaction medium, while homogeneous catalysts are soluble.

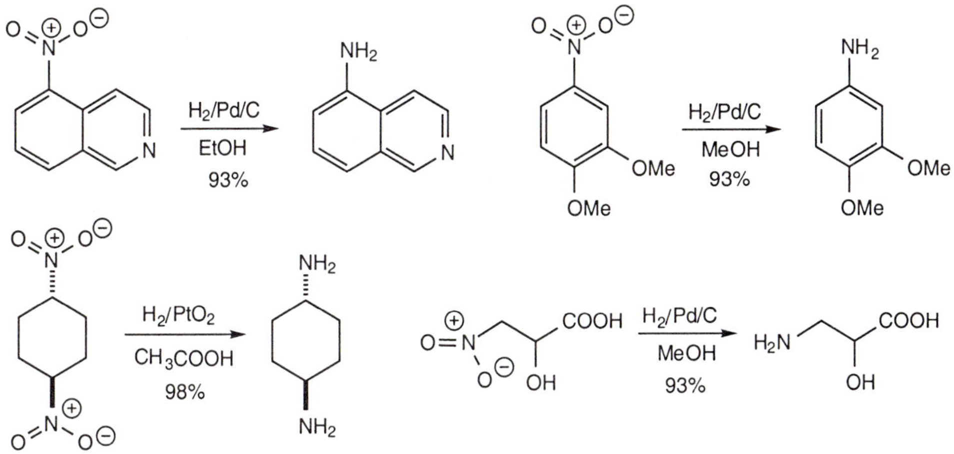

romatic nitro compounds (such as nitrobenzene) play a pivotal role in the ynthesis of substituted arenes partly because they direct electrophilic romatic substitution so powerfully. Also remember that one can substitute a itro group *via* reduction to an amine and formation of a diazonium salt with HNO_2. It is useful, therefore, to be able to reduce a nitro group to an amine nd heterogeneous hydrogenation with a transition metal catalyst (usually latinum or palladium) is best in this regard, Fig. 6.1.

H_2/Pd/C, EtOH, 93%

H_2/Pd/C, MeOH, 93%

H_2/PtO_2, CH_3COOH, 98%

H_2/Pd/C, MeOH, 93%

Fig. 6.1

In general terms, both aromatic and aliphatic nitro groups are easy to reduce, often without affecting other functionality.

We will discuss catalytic hydrogenation later on (Chapter 7) but at this point it is sufficient to note that hydrogen gas is dissociatively adsorbed onto the surface of the metal. Subsequent adsorption of the nitro compound onto the metal is followed by addition of two hydrogen atoms to the nitro group; this process is repeated until the fully reduced amine is formed, Fig. 6.2.

Sometimes it is possible to isolate the partially reduced intermediates that are shown in Fig. 6.2; however, specialised reducing agents are required if the partial reduction is to be of preparative value.

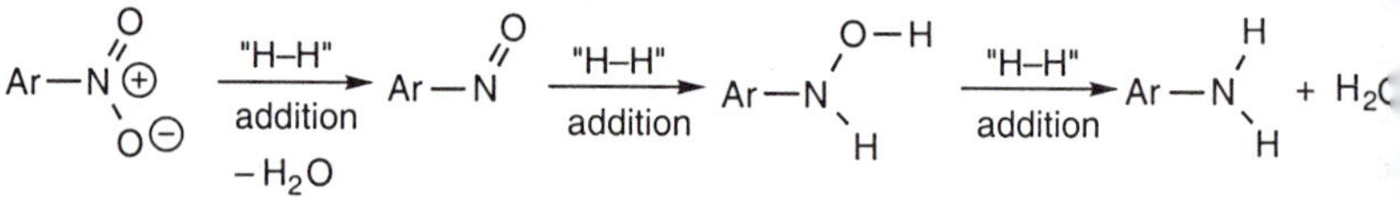

Fig. 6.2

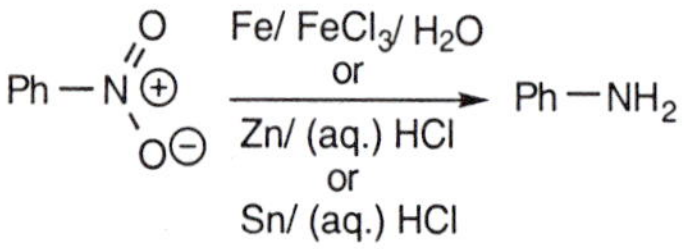

Reactive metals (such as tin or iron or zinc) comprise another useful class of reducing agents that find application in the transformation of nitro groups to amines.

The mechanism of this reduction is presumed to involve stepwise addition of electrons (these are provided by the metal, for example, Zn→ Zn^{2+} + 2e) and subsequent protonation of the charged intermediates by acid, Fig. 6.3.

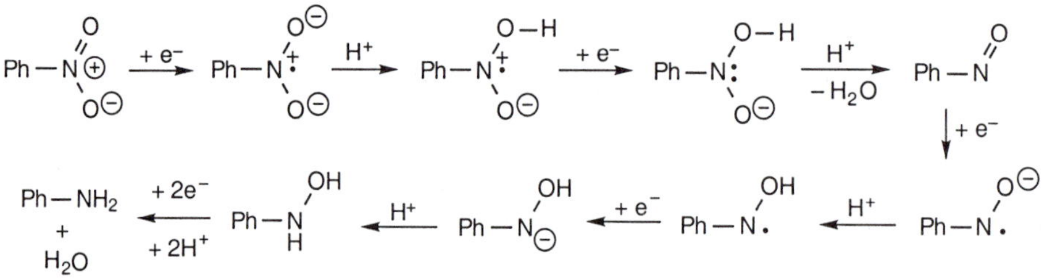

Fig. 6.3

Reduction of amine-*N*-oxides to amines

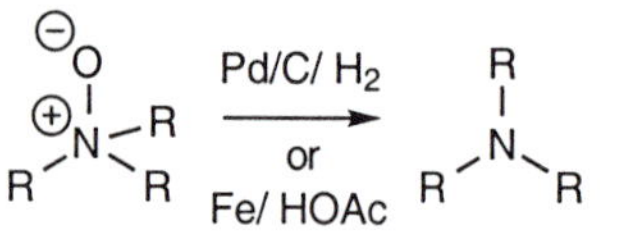

It is possible to use catalytic hydrogenation (or reactive metals in acid) to deoxygenate amine-*N*-oxides to the corresponding amines. The *N*-oxide of pyridine is important in synthesis because it activates the pyridine ring to electrophilic aromatic substitution at the C-4 position; it is important, therefore, that the *N*-oxide can be reduced at the end of a synthetic sequence. The example shown in Fig. 6.4 involves the reduction of both an amine-*N*-oxide and a nitro group at the same time.

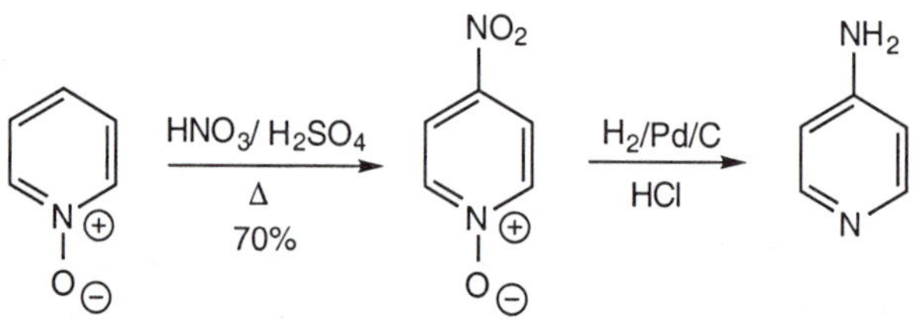

Fig. 6.4

eduction of azides (RN_3) to amines (RNH_2)

ne azide ion (N_3^-) is an incredibly good nucleophile and this may be tributed to its small size, negative charge, and, being a stabilised anion, ck of basicity. If one wishes to introduce nitrogen into an organic mpound *via* a displacement reaction then the use of N_3^- in an S_N2 reaction very effective. Quite often it is useful to be able to reduce the azide so rmed to an amine and several methods have been developed, Fig. 6.5. erhaps the most useful is reduction with hydrogen using a palladium or atinum catalyst, which gives very high yields of amine. As you will see ter, such hydrogenation conditions will also reduce alkenes to alkanes very ell indeed. A nice solution to the problem of reducing azides to primary nines in the presence of an alkene has been found which involves a modified talyst that does not reduce alkenes at all (we shall cover this catalyst, nown as Lindlar's catalyst, later on).

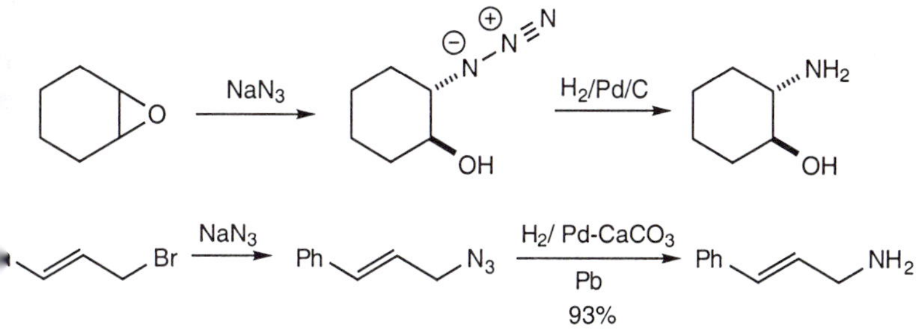

Lindlar's catalyst consists of palladium on a calcium carbonate support. Lead is added to deactivate (poison) the catalyst and make it ineffective at reducing alkenes. Amine additives such as quinoline also aid in poisoning the catalyst.

Fig. 6.5

Another method of reducing azides that is particularly useful involves action with triphenylphosphine and water, Fig. 6.6. Initial attack of the osphine onto the azide yields a phosphazide **1** which then decomposes (*via* ss of nitrogen) to a phosphine imide. Subsequent hydrolysis of the osphine imide gives an amine and triphenylphosphine oxide (so, in this quence the azide has been reduced and the triphenylphosphine oxidised).

As you might expect, this reducing method does not affect carbon–carbon double bonds.

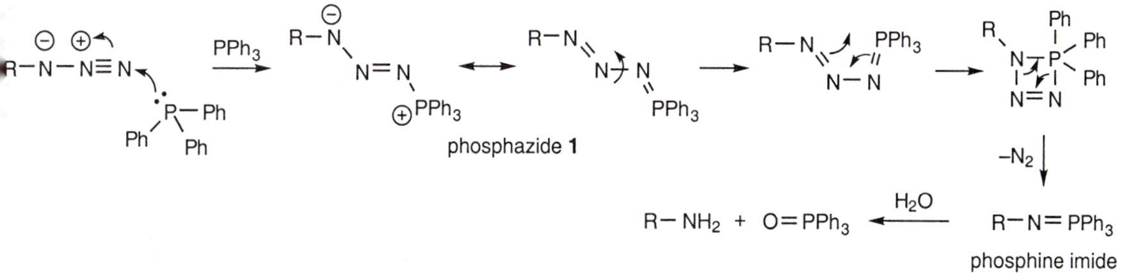

Fig. 6.6

eduction of N–O bonds in oximes and hydroxylamines

e N–O single bond in both oximes and hydroxylamines can be cleaved ductively with hydrogen and a metal catalyst (platinum is best). These are eful reactions because the starting materials can be made *via* the cloaddition of nitrile-*N*-oxides or nitrones to alkenes.

Nitrile-*N*-oxides are normally formed *in situ* by dehydration of a nitro compound or elimination from an α-halo oxime.

For example, reaction of nitrile-*N*-oxide **2** gave heterocycle **3** which w then reduced, Fig. 6.7. Presumably, addition of hydrogen across the N- bond gave a hydroxy imine **4** which was hydrolysed to the ketone *in sit* This chemistry was developed as a potential route to the natural produ gephyrotoxin.

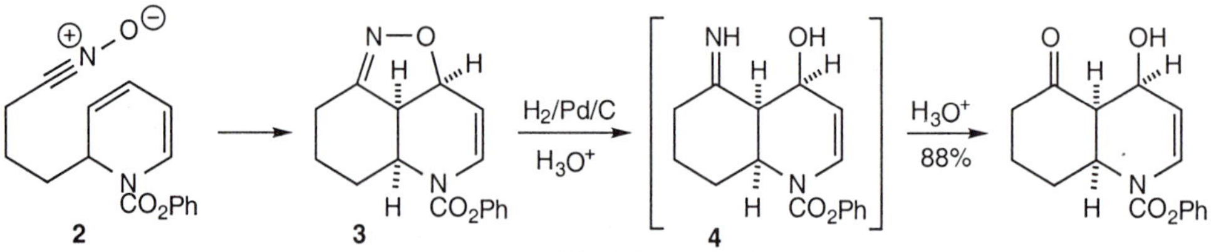

Fig. 6.7

In a related sequence, the nitrone **5** underwent cycloaddition with electron-deficient alkene to yield a cycloadduct **6**, Fig. 6.8. After reacti with methanesulfonyl chloride, the N–O bond in **6** was cleaved under standa conditions. The hydroxy-amine product from reduction then cyclised (v intramolecular displacement of the OSO_2Me group) to a bicyclo compou that was eventually converted into the alkaloid supinidine.

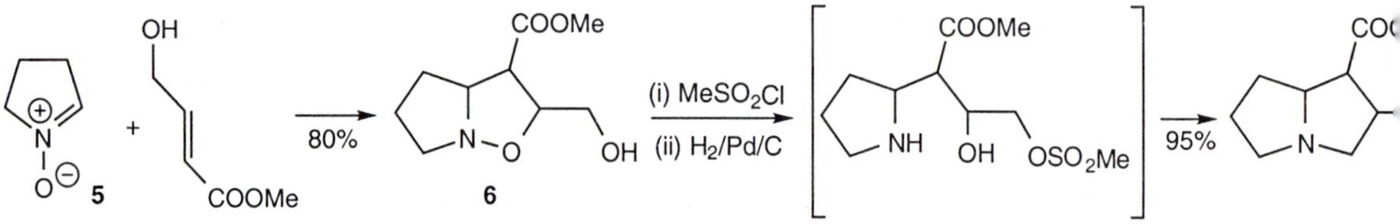

Fig. 6.8

Reduction of N–O bonds in biology

Compound **7**, known as FR66979, is a potent antitumour antibiotic which capable of crosslinking strands of DNA, Fig. 6.9. Studies on the mechanis of action of this compound have revealed that it requires reduction of the N- bond *in vivo* in order to display its biological activity. So, reductive cleava of the N–O bond of FR66979 leads to a rearrangement process that eventual produces compound **8**, which is a potent electrophile capable of reacting wi the bases of DNA. Although the mechanism of bioreduction is not known could involve a metal-ion-containing enzyme.

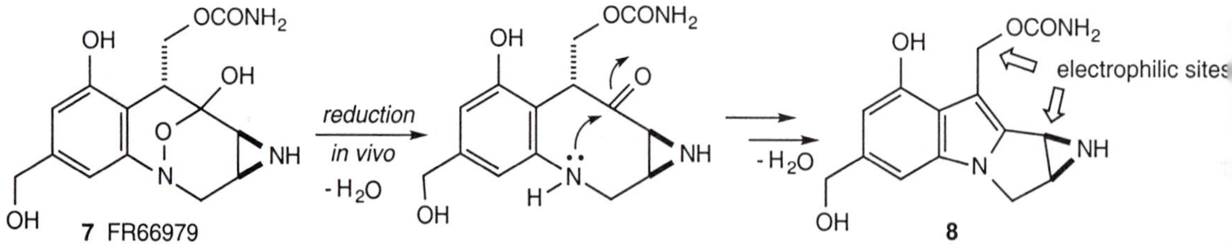

Fig. 6.9

Cleavage of N–S bonds: reduction of sulfonamides

The N–S bond of a sulfonamide may be reductively cleaved with reactive metals in the presence of a proton source, Fig. 6.10. Typical conditions are sodium in liquid ammonia or sodium amalgam in buffered methanol. As this process liberates the parent amine, sulfonamides are often used as protecting groups for secondary amines.

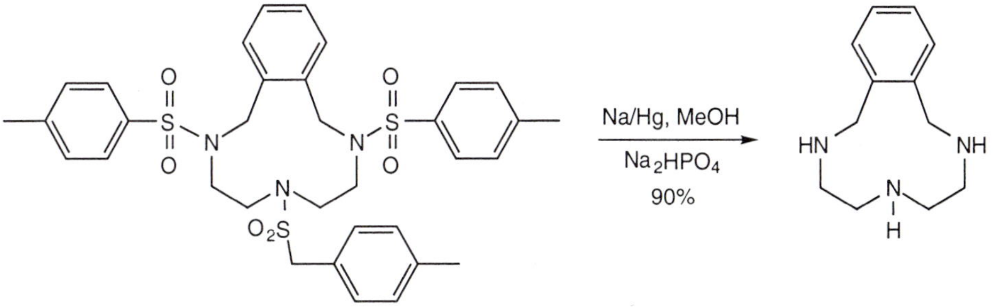

Na/Hg, MeOH; Na_2HPO_4; 90%

Fig. 6.10

The suitability of sulfonamides as protecting groups is enhanced by the observations that they are normally stable to basic hydrolysis and relatively inert to hydrogenation. So, selective deprotection of other functional groups can be achieved in the presence of these derivatives.

A plausible mechanism which explains the reduction involves addition of an electron to the sulfonamide to form a radical anion which may cleave to yield a nitrogen-based anion and a sulfur-based radical (this will then be reduced again), Fig. 6.11. Subsequent protonation of the nitrogen anion *in situ* forms the corresponding amine.

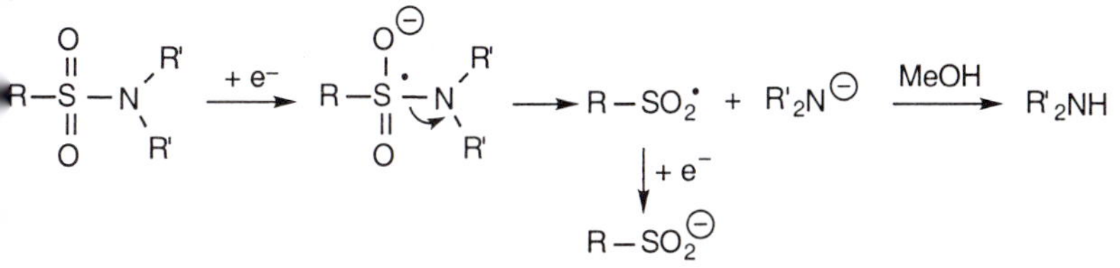

$$R{-}SO_2{-}NR'_2 \xrightarrow{+e^-} [R{-}SO_2{-}NR'_2]^{\bullet -} \longrightarrow R{-}SO_2^{\bullet} + R'_2N^{\ominus} \xrightarrow{MeOH} R'_2NH$$

$$R{-}SO_2^{\bullet} \xrightarrow{+e^-} R{-}SO_2^{\ominus}$$

Fig. 6.11

6.2 Reduction of phosphorus–oxygen bonds

Reduction of a phosphorus–oxygen double bond (for example, reducing a phosphine oxide to a phosphine) is not a particularly common reaction, and there is not a great deal of work in this area. Remember that the P=O bond is very strong, which does not help matters. However, reducing agents such as $LiAlH_4$ and, in particular, alane (AlH_3) have recently been shown to be effective for this transformation, Fig. 6.12.

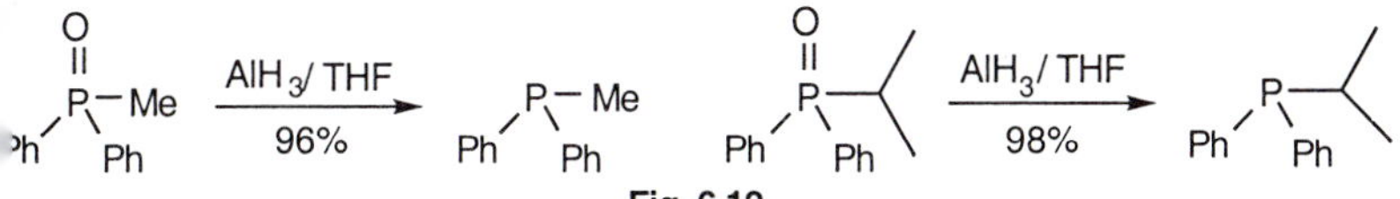

$$Ph_2P(O)Me \xrightarrow[96\%]{AlH_3/THF} Ph_2PMe \qquad Ph_2P(O)CH(CH_3)_2 \xrightarrow[98\%]{AlH_3/THF} Ph_2PCH(CH_3)_2$$

Fig. 6.12

Alane is made by adding concentrated H_2SO_4 to $LiAlH_4$.

The mechanism of this reaction has not been studied in detail but coordination of the Lewis acidic AlH_3 to the oxygen of the P=O to give intermediate **9** seems reasonable. Nucleophilic attack by hydride onto the phosphorus atom could then give an intermediate capable of undergoing an α-elimination to the phosphine, Fig. 6.13.

Alane, just like borane, is an electron-deficient Lewis acidic complex. Once a Lewis base has coordinated to alane then it becomes a negatively charged ATE complex (e.g. aluminate complex **9**) that can expel H^-, and so act as a reducing agent.

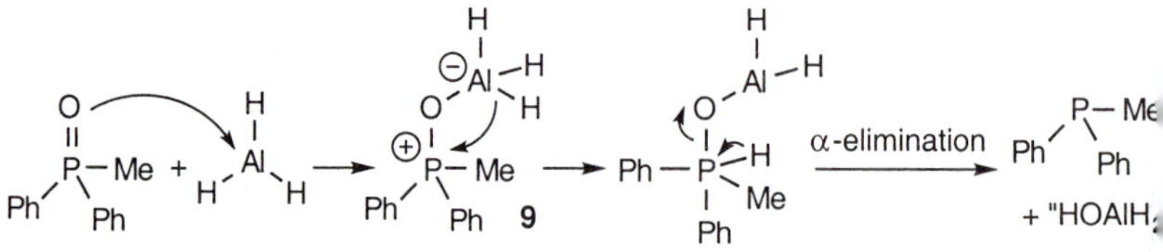

Fig. 6.13

6.3 Reduction of sulfur– and selenium–containing functional groups

Reduction of disulfides and diselenides

Disulfides (RSSR) and diselenides (RSeSeR) contain a weak and polarisable heteroatom–heteroatom bond and consequently are easy to cleave to the corresponding thiol (RSH) and selenol (RSeH) respectively. Sodium borohydride ($NaBH_4$) is one reducing agent that will cleave both efficiently Fig. 6.14.

Sodium borohydride is a mild reducing agent that behaves as a nucleophilic source of H^-. In this reaction the BH_4^- ion probably delivers H^- to the disulfide displacing a RS^- group: this ion can pick up a proton during work-up.

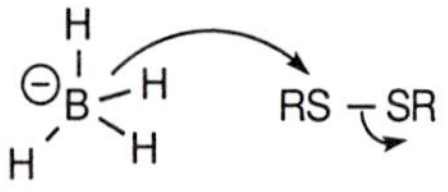

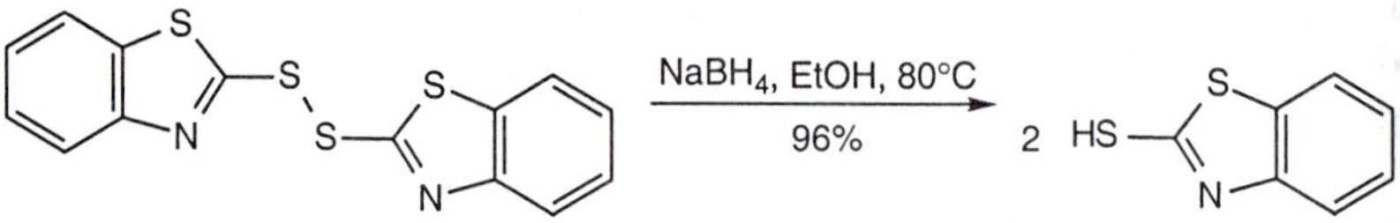

Fig. 6.14

Reduction of diselenides using this method generates the corresponding anion (RSeNa) which can be used as a nucleophile *in situ* (Fig. 6.15 shows PhSeNa opening an epoxide).

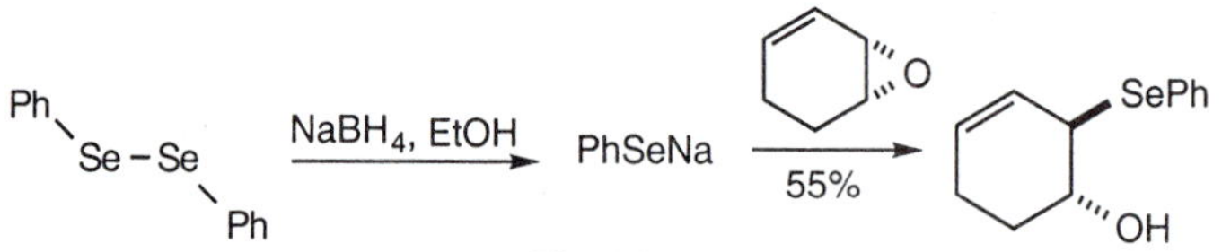

Fig. 6.15

Reduction (deoxygenation) of sulfoxides

There are many reagents which are capable of deoxygenating sulfoxides to yield sulfides ($TiCl_3$, VCl_2, AlH_3, H_2/Pd-C); we shall only consider in detail here reduction of sulfoxides with phosphites [$(RO)_3P$], Fig. 6.16. All sorts of R groups can be tolerated within the structure of the phosphite but R=Ph is an effective reagent as well as the one shown in Fig. 6.16. Presumably, the driving force for this reduction is the fact that the P=O bond is stronger than the S=O bond.

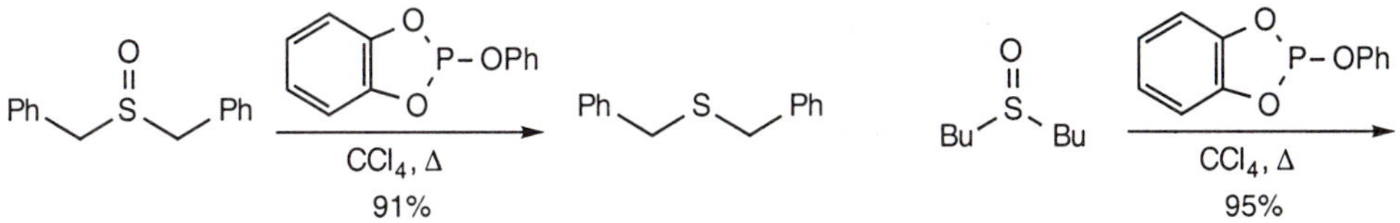

Fig. 6.16

Although we could draw other mechanisms, initiation by nucleophilic attack of phosphorus onto sulfur to form a three-membered ring is shown. Collapse of this intermediate yields the sulfide and phosphorus compound, Fig. 6.17. Clearly, this sequence has led to the reduction of sulfur and the oxidation of phosphorus.

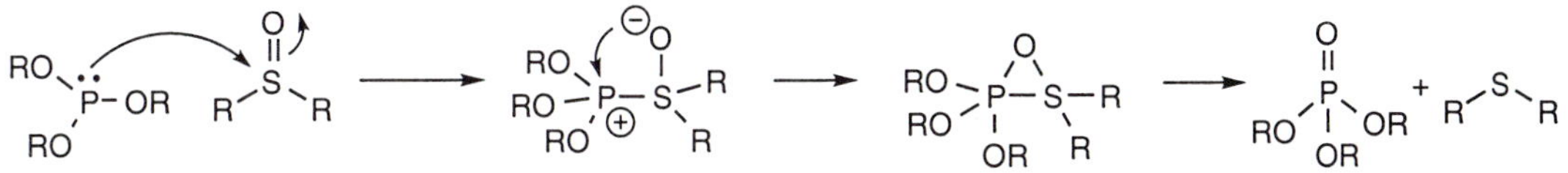

Fig. 6. 17

Reduction (deoxygenation) of sulfones

The literature is sparse in this area and there is not a really good and general reagent to effect the reduction of sulfones to either sulfoxides or sulfides. Lithium aluminium hydride ($LiAlH_4$) gives moderate yields of sulfides while diisobutylaluminium hydride (i-Bu_2AlH or DIBAL-H) is probably the best reducing agent known for such a transformation, Fig. 6.18.

Reduction of sulfones with $LiAlH_4$ may involve nucleophilic attack of H^- onto the sulfur atom, followed by an α-elimination (compare with Fig. 6.13); this would explain why di-*t*-butyl sulfone is inert to $LiAlH_4$ as the sulfur atom is too hindered (formally a neopentyl centre) to act as a good electrophile.

$LiAlH_4$ is another source of nucleophilic H^-, although it is much more reactive than sodium borohydride and consequently is used in aprotic solvents such as THF rather than ethanol.

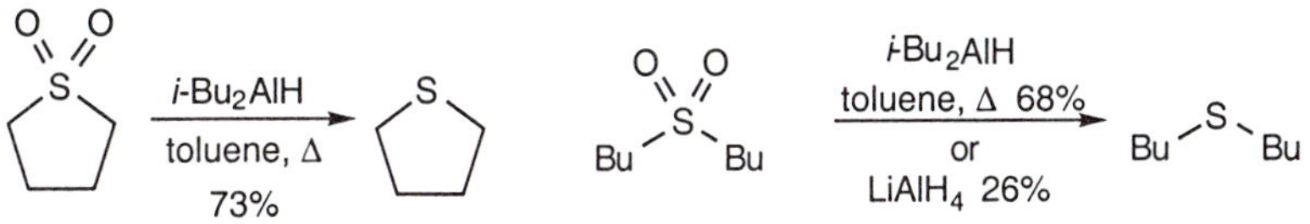

Fig. 6.18

7. Reduction of carbon–carbon double and triple bonds

Naturally, most of the functional groups that concern us as organic chemists are based on carbon. In this chapter we shall examine ways of reducing multiple (π) bonds between carbon atoms by the addition of hydrogen (a process known as hydrogenation).

7.1 Reduction of alkenes

Heterogeneous hydrogenation

The use of a transition metal catalyst to promote addition of hydrogen to an organic compound can be traced back to experiments performed by Sabatier in 1897. Modern procedures involve a transition metal catalyst (Rh, Pt, Pd are the most common) adsorbed onto a solid support (such as carbon or alumina). Although elevated pressures and temperatures invariably increase the rate of such hydrogenations, the reactions usually proceed in a satisfactory manner at room temperature under one atmosphere of H_2.

Remember that hetereogeneous catalysts are insoluble in the reaction medium.

So, reaction of an alkene with hydrogen gas and one of the above catalysts results in reduction of the π-bond and formation of an alkene, Fig. 7.1. The stereospecific reduction of *E*- and *Z*-**1** shows that the addition of hydrogen occurs in a *syn* fashion.

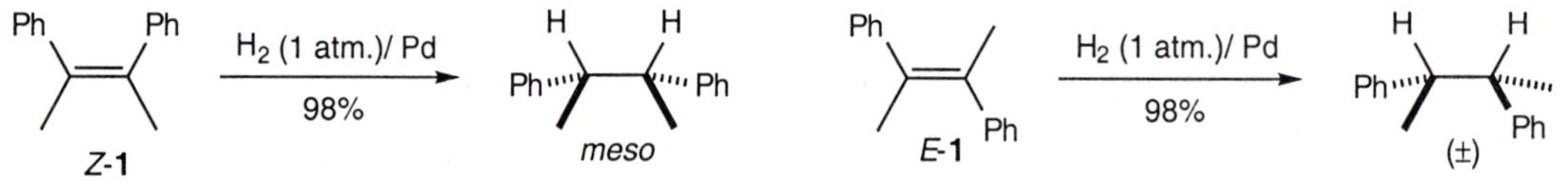

Ph Ph *Z*-**1** → H_2 (1 atm.)/ Pd, 98% → H H Ph Ph *meso*

Ph Ph *E*-**1** → H_2 (1 atm.)/ Pd, 98% → H H Ph Ph (±)

Fig. 7.1

However, the amount of *syn*-addition that is observed is usually somewhat less than complete, as witnessed by low selectivity during the reduction of **2**, Fig. 7.2.

It is interesting to note that the selectivity for *cis*-**3** was improved to 96:4 using Pt catalysis and 300 atmospheres of hydrogen gas. Levels as high as 98:2 were obtained with osmium or iridium catalysts.

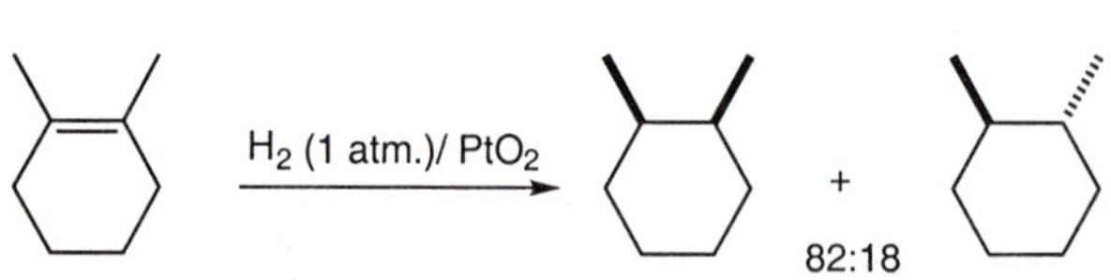

2 → H_2 (1 atm.)/ PtO_2 → *cis*-**3** + (±)–*trans*-**3** (82:18)

Fig. 7.2

The mechanism by which hetereogeneous hydrogenation proceeds is complex and difficult to study as the reaction takes place on the surface of the metal (and the nature of this surface varies with the different catalysts used). A working model for the hydrogenation process assumes that first H_2 is dissociatively adsorbed onto the catalyst surface, Fig. 7.3. Alkene π-bonds may also be adsorbed onto the metal, forming complexes which are anchored at two points. Subsequent addition of a hydrogen atom to one of the carbon atoms of the adsorbed species leaves an intermediate which is referred to as a half-hydrogenated state **4**. All that remains is addition of a second atom of hydrogen onto the other carbon atom and the reduced product then can dissociate from the catalyst. As one face of the alkene is adsorbed onto the metal and both hydrogen atoms then transfer from the metal to the π-system, this model explains why the addition of H_2 is *syn*.

Extensive studies of the rates of hydrogenation of alkenes reveal that highly substituted alkenes are reduced more slowly than less substituted alkenes ($R_2C{=}CR_2$ < $RCH{=}CR_2$ < $RCH{=}CHR$ < $RCH{=}CH_2$). This ordering may be related to the observation that steric hinderance around the π-bond hinders adsorption onto the metal surface.

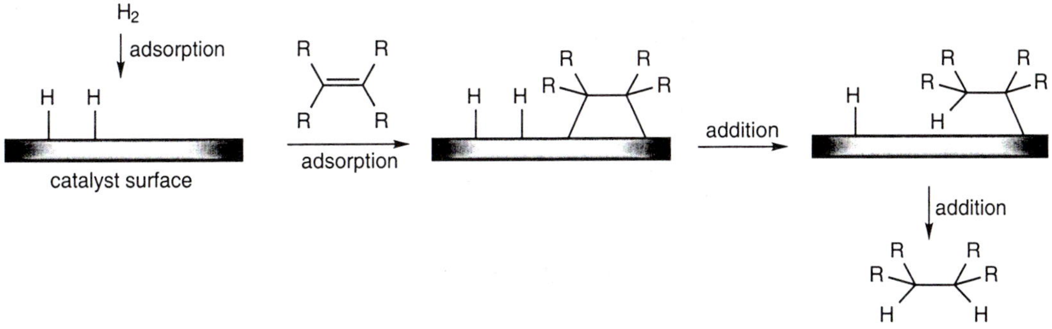

H2
adsorption
H H
catalyst surface
adsorption
addition
4
addition

Fig. 7.3

Now, the unexpected phenomenon regarding *apparent* lack of complete *syn*-addition of hydrogen (see Fig. 7.2) may be explained if we suppose that all steps except for the last of this sequence are reversible. So, the half-hydrogenated state may revert back to the fully adsorbed state as well as adding another atom of hydrogen to become product. However, if one of the R groups has a suitably positioned hydrogen atom then this reversion need not occur in the direction from which it came, Fig. 7.4. Consider the half-hydrogenated state of compound **2** (from Fig. 7.2); this may revert back to the original compound or, alternatively, to two other alkenes **5** and **6**. Of course, **5** and **6** will then be hydrogenated under the reaction conditions, but there is no reason why they should give exclusively *cis*-**3** as products. This process of alkene migration is common under hydrogenation conditions and can often go unnoticed especially when each isomer of the starting material gives the same product on reduction.

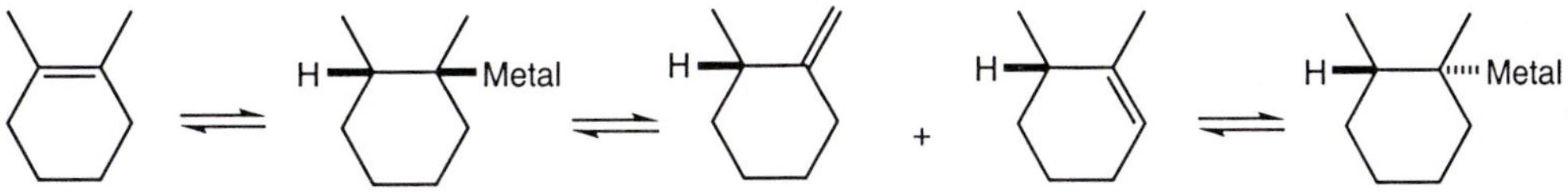

2
half-hydrogenated state
Metal
5
6
Metal

Fig. 7.4

The most commonly used solvents for hydrogenation are methanol and ethanol, which are capable of dissolving sufficient amounts of hydrogen.

Adding more hydrogen to the metal surface is an obvious tactic for reducing double-bond migration with respect to hydrogenation. As hydrogen becomes more abundant on the metal surface, the half-hydrogenated state will add a second atom of hydrogen more rapidly than it reverts back to the fully adsorbed state. This explains why increasing the hydrogen pressure for the reaction shown in Fig. 7.2 increases the *cis* selectivity by reducing the amount of double-bond migration.

Homogeneous hydrogenation

Wilkinson's catalyst ($[Ph_3P]_3RhCl$) **7** is a most useful (soluble) metal complex for the homogeneous hydrogenation of alkenes. So, addition of hydrogen to a red solution of **7** promotes the formation of (yellow) complex **8**, by oxidative addition of hydrogen and dissociation of a bulky phosphine ligand from the metal, Fig. 7.5. In fact, dissociation of a phosphine ligand opens up a coordination site for the alkene to attach itself to the metal. The next step in the mechanism involves the concerted addition of a metal–hydrogen bond across the coordinated alkene to form complex **9** (both atoms are added in a *cis* fashion to the alkene). Finally, reductive elimination ensues, forming a second carbon–hydrogen bond and generating **10** which adds hydrogen to become **8** and so continues the catalytic cycle.

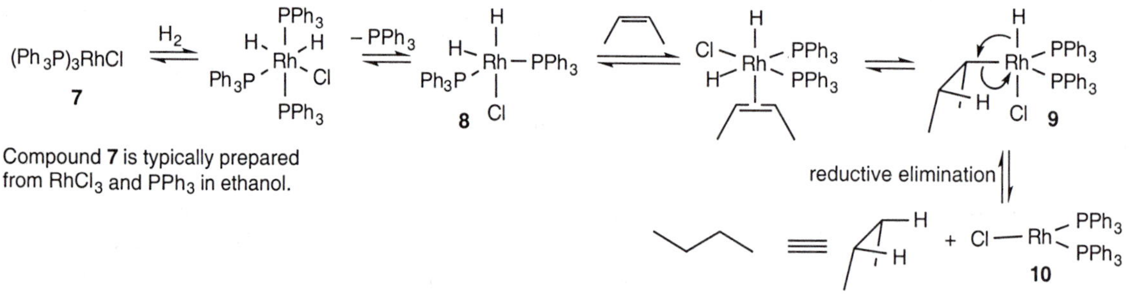

Compound **7** is typically prepared from $RhCl_3$ and PPh_3 in ethanol.

Fig. 7.5

Sterically hindered alkenes are reduced more slowly with this catalyst and, unlike homogeneous hydrogenation, scrambling of the double bond does not occur during the reduction process. These properties make Wilkinson's catalyst ideal for the selective hydrogenation of unhindered alkenes, Fig. 7.6.

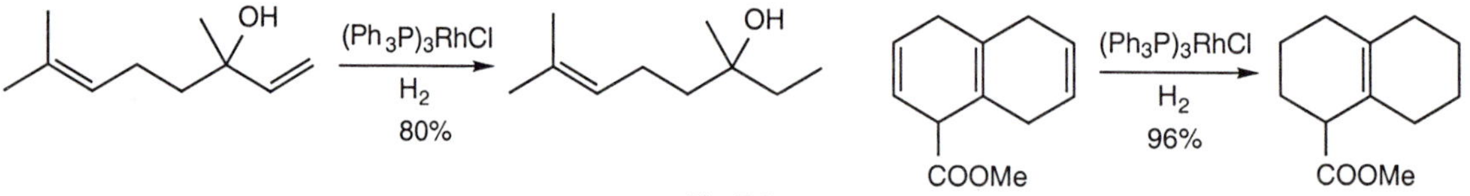

Fig. 7.6

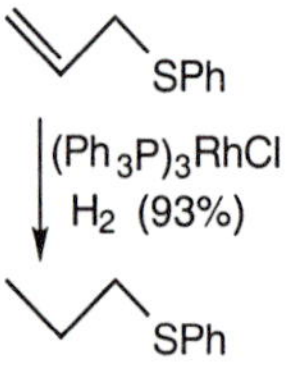

Divalent sulfur-containing functional groups normally deactivate heterogeneous transition metal catalysts, thus preventing hydrogenation. However, this is not the case for homogeneous hydrogenation and Wilkinson's catalyst can bring about the reduction of allyl phenyl sulfide smoothly.

Stereoselectivity in the hydrogenation of alkenes

Consider hydrogenation of alkene **11**: under normal homogeneous or heterogeneous hydrogenation conditions addition of hydrogen to both faces of the alkene will occur in equal proportions and the product will be racemic. Now, if it were possible to incorporate a chiral (non-racemic) ligand into the hydrogenation mixture then, in principle, discrimination between the two (enantiotopic) faces could take place and the product form as a single enantiomer.

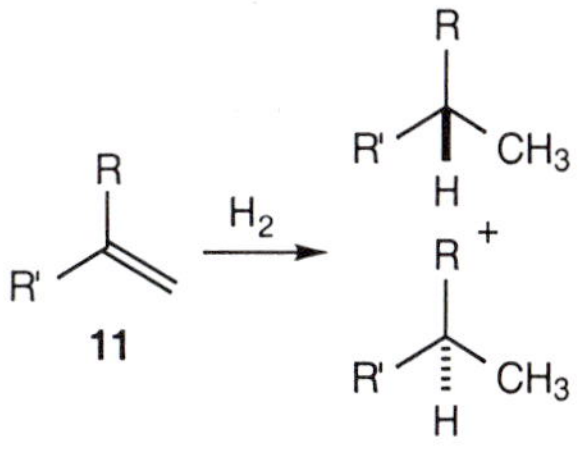

The mechanism of homogeneous hydrogenation is easier to study than that of heterogeneous reduction and most effort has gone into developing chiral ligands (normally replacements for triphenylphosphine) for this process. Both rhodium and ruthenium metals work rather well, and a general-purpose chiral ligand is BINAP (commercially available as either enantiomer).

The facial selectivity of hydrogenation has its origins in the formation of a complex between the metal, BINAP and alkene. Complexes with each of the two faces of the alkene are diastereoisomeric by virtue of the chiral ligand and it is hydrogenation of these diastereoisomeric complexes at different rates that determines the stereoselectivity of the product. Although the precise details by which the chiral ligand exerts its influence have been studied in detail, they will not be covered here.

So, reduction of the alkene **12** using catalytic Rh(I) and catalytic (*S*)-BINAP gives the product as essentially a single enantiomer, Fig. 7.7. It should be noted, however, that not all types of alkene can be reduced successfully with this catalyst and ideally the alkene should have both an electron-withdrawing group and a group with a chelating carbonyl (OAc, NHAc, etc.) on one of the alkene carbons.

BINAP is a bidentate ligand for transition metals. It is chiral by virtue of the fact that the rings do not lie coplanar; it has axial chirality.

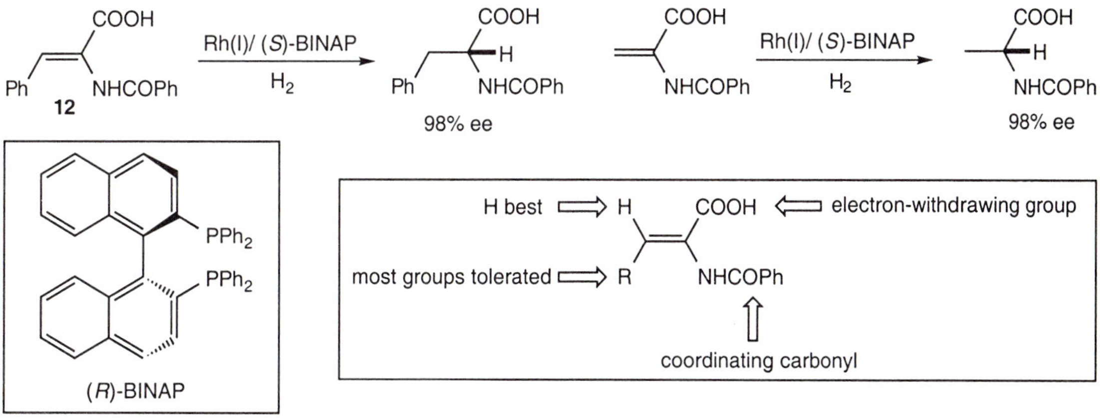

Fig. 7.7

Further development of the hydrogenation process led to the use of chiral ruthenium(II) catalysts and, again, BINAP is effective as a chiral ligand, Fig. 7.8 (although there are many different types of chiral phosphine ligand available and the best one for any particular alkene is found by experiment).

Ruthenium catalysts reduce a variety of alkenes with high enantioselectivity and are therefore more generally applicable than rhodium catalysts.

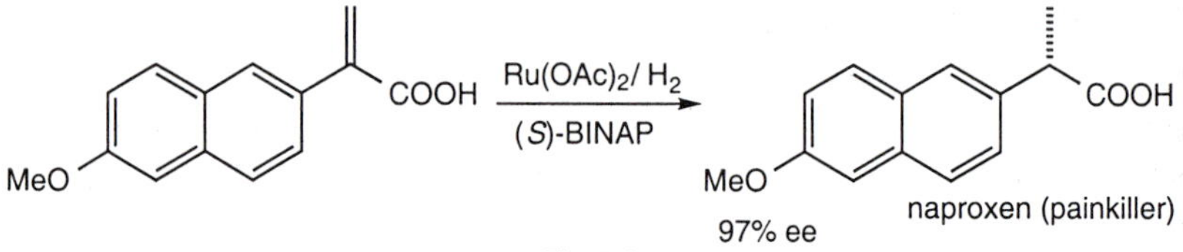

MeO COOH $Ru(OAc)_2/H_2$ (*S*)-BINAP → MeO COOH naproxen (painkiller) 97% ee

Fig. 7.8

Reduction of electron-deficient alkenes with metals in ammonia

Solutions of group I metals in ammonia are blue; this is the colour of the solvated electron formed by the process, $Na \rightarrow Na^+ + e^-$. You can consider these solutions as sources of free electrons.

Alkenes that are substituted with an electron-withdrawing group can be reduced with lithium (or sodium or potassium) in liquid ammonia at low temperature with the addition of one equivalent of a proton source, normally an alcohol. Alkenes that are not substituted with an electron-withdrawing group are unreactive and are left untouched, Fig. 7.9.

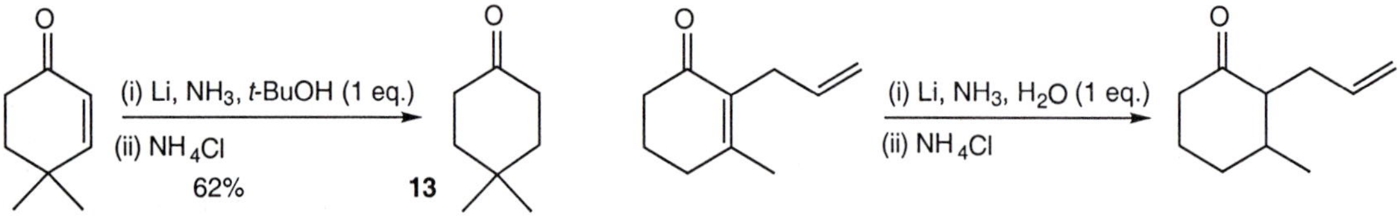

(i) Li, NH_3, *t*-BuOH (1 eq.) (ii) NH_4Cl 62% → **13**

(i) Li, NH_3, H_2O (1 eq.) (ii) NH_4Cl →

Fig. 7.9

Ammonia is a gas at room temperature and condenses at –33°C.

An alternative pathway for the formation of **14** involves direct protonation of the radical anion on carbon, followed by addition of a second electron.

Addition of an electron to an alkene π-system results in formation of a radical anion, Fig. 7.10 (the more electron deficient the π-system the easier it is to add an electron). The radical anion may then be protonated by the alcohol to yield a neutral radical which subsequently adds another electron to form an allylic anion. Proton transfer ensues and the result is formation of an enolate species **14**. Under normal conditions **14** is stable and unable to gain a proton from ammonia (NH_3, $pK_a \approx 34$: remember that there is only one equivalent of alcohol added). Addition of a more acidic proton source, such as NH_4Cl, after the reaction is complete, protonates the enolate and forms the carbonyl compound **13**.

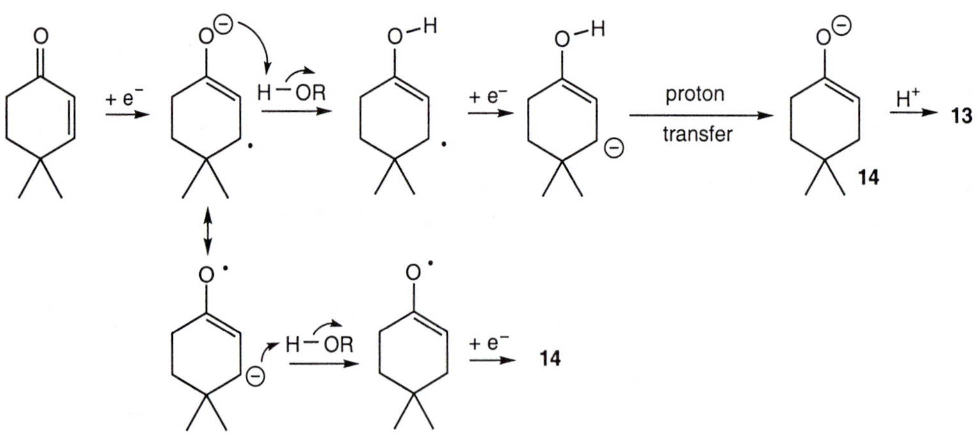

$+e^-$ → H–OR → $+e^-$ → proton transfer → **14** H^+ → **13**

H–OR → $+e^-$ → **14**

Fig. 7.10

7.2 Reduction of alkynes

Alkynes may be reduced to alkanes by the addition of two moles of hydrogen and this reaction is best accomplished by heterogeneous hydrogenation with palladium or homogeneous hydrogenation with $(Ph_3P)_3RhCl$. As one might expect, the reduction proceeds *via* an alkene which is reduced further *in situ*.

This leads to a rather interesting problem: how can we accomplish the partial reduction of alkynes to alkenes? Any solution that we find should be capable of controlling the geometry of the alkene formed so that we can prepare either *cis*- or *trans*- alkenes from any particular alkyne.

R
R
H_2, Pd-C
or
H_2, $(PhP)_3RhCl$
R
R

Formation of *cis*-alkenes from alkynes

Perhaps the best known method for the partial reduction of alkynes to alkenes is heterogeneous hydrogenation with Lindlar's catalyst (which we have seen before in Chapter 6 as a means of reducing azides). Lindlar's catalyst is essentially palladium that has been poisoned with lead and an amine (typically quinoline) so that it is ineffective at reducing alkenes.

Bearing in mind that the mechanism for reduction with Lindlar's catalyst is similar to that shown in Fig. 7.3, it should come as no surprise to note that the two hydrogens are added to the alkyne in a *cis* fashion, and the result is formation of a *cis*-(Z)-alkene **15**, Fig. 7.11.

Amine additives deactivate the catalyst by binding onto active sites on the metal surface. This binding is in direct competition with both the alkyne and alkene product, and it reduces the catalyst's effectiveness. The additives may also aid rearrangement of the catalyst surface to a less active morphology.

( )6
O
OH
Pd, $BaSO_4$
Pb, quinoline
H_2
H
H
O
OH
( )6
Z-**15** 90%

Fig. 7.11

On occasion, problems such as over-reduction and a lack of *cis* stereoselectivity can arise with Lindlar hydrogenation; in this author's experience, nickel boride (made from $Ni(OAc)_2$ and sodium borohydride) is an excellent alternative catalyst for the synthesis of *cis*-alkenes from alkynes.

Hydroboration of alkynes

Diborane (B_2H_6) is another reducing agent that may be used to produce *cis*-alkenes from alkynes. Concerted addition of an H_2B–H bond across the alkyne π-system **16** results in formation of a vinyl borane, Fig. 7.12. This addition occurs twice more until each of the B–H bonds has been replaced by a B–C bond. In a second step the vinyl borane **17** is treated with acetic acid at low temperature and the carbon–boron bond is replaced (with retention of configuration) by a carbon–hydrogen bond to form a *cis*-alkene with high stereoselectivity.

You can consider the hydroboration reaction itself as a reduction, followed by a reaction that is simply replacement of one electropositive element (B) with another (H).

Et —≡— Et
16
B_2H_6
[H, BH_2, Et, Et]
vinyl borane
2 x **16**
(H, B, Et, Et)$_3$
17
HOAc, 0°C
3 x H, H, Et, Et
68% ≥ 98% *cis*

Fig. 7.12

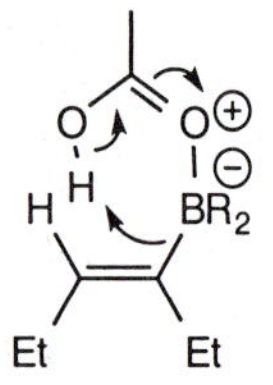

The mechanism for the second step is presumed to involve prior coordination of acetic acid to the electron-deficient boron, followed by migration of carbon to hydrogen as shown.

Formation of *trans*-alkenes from alkynes

Formation of *trans*-(*E*)-alkenes from alkynes can be quite tricky to accomplish. Sodium in liquid ammonia at low temperature is one method of achieving this transformation with high levels of stereoselectivity (however, be aware that this reagent is capable of reducing many other types of functional group).

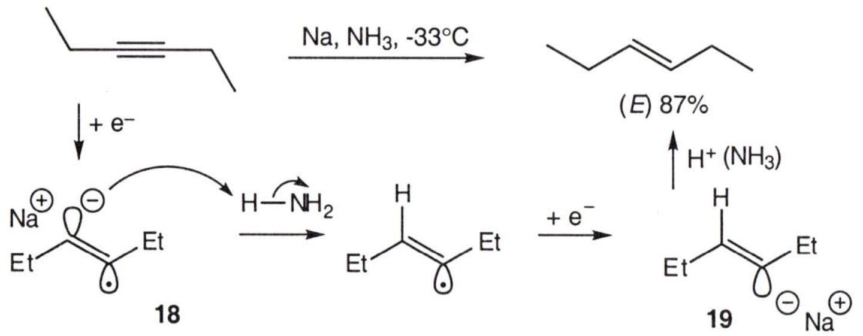

Na, NH_3, -33°C

(*E*) 87%

+ e⁻

H^+ (NH_3)

$H{-}NH_2$

+ e⁻

18

19

Fig. 7.13

The mechanism of the reduction is thought to proceed *via* addition of an electron to the alkyne to form a radical anion **18** (which adopts *trans* geometry), Fig. 7.13. Protonation of the radical anion ensues (the proton can come from ammonia) and the resulting radical picks up another electron to give the vinyl anion **19**. Finally, protonation of this anion leads to the observed *trans*-alkene. Fortunately, it is more difficult to reduce alkenes than alkynes under these conditions and so the over-reduction is not normally observed.

7.3 Reduction of aromatic π-systems

Partial reduction of aromatic compounds

The Birch reduction is particularly useful and unique in its ability to accomplish the partial reduction of aromatic compounds. Electrons are the reducing agent and again these come from the group I metals sodium, lithium or potassium in liquid ammonia (all are accompanied by a characteristic blue colour). So, reduction of benzene with sodium, ammonia and *t*-BuOH at low temperature (–78°C) will form the dihydro-product in good yield, Fig. 7.14. The two alkene units in the reduced compound are not in conjugation with each other, which appears a little odd at first. The mechanism for reduction is similar to that for the partial reduction of alkynes and enones and involves sequential addition of an electron, followed by protonation of the radical anion species so formed.

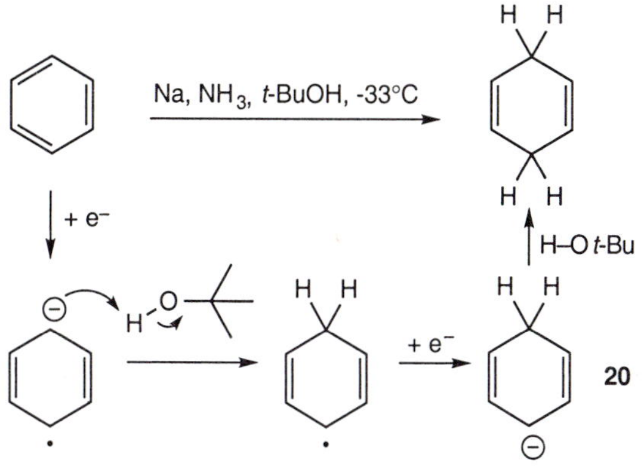

The non-conjugated position of the two alkenes in the product is determined by the regioselectivity of protonation of pentadienyl anion **20**. Only protonation at the middle carbon will do, as protonation at either end would give the conjugated product. Why do pentadienyl anions protonate faster at the middle carbon than at the ends? This is a difficult question to answer intuitively, but calculations have shown that the central carbon has a higher concentration of electron density than the end carbons. So, *kinetic* protonation at the most electron rich site gives the observed regioselectivity.

Fig. 7.14

The situation becomes more complicated when substituted aromatic ompounds are reduced under Birch conditions, Fig. 7.15. As you can see, lectron-rich compounds such as anisole and toluene give products where the ubstituent is conjugated with one alkene.

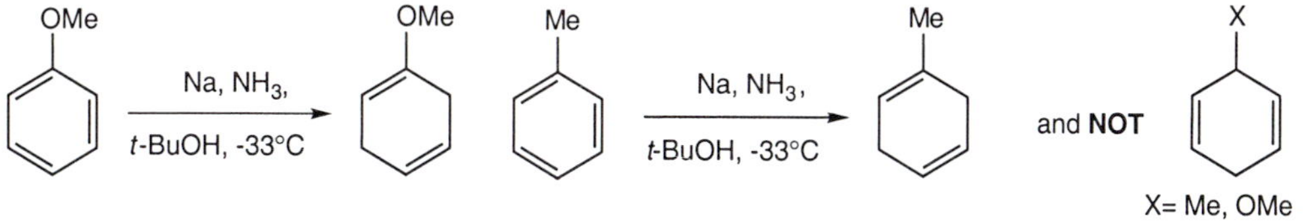

Fig. 7.15

The regiochemistry observed in the Birch reduction of substituted aromatic ompounds has its origins in the protonation of the radical anion formed after ddition of a single electron to the aromatic π-system. Factors to think about uring the protonation of such radical anions are (i) which carbon is the site f highest electron density; (ii) what is the stability of the radical formed pon protonation? Consider **21**, Fig. 7.16, which must protonate referentially at C-2 rather than at C-1 or C-4 to give the observed products. rotonation takes place at the carbon with the highest electron density and an be explained simply by considering that canonical **21b** is a truer picture f reality than **21a** because a negative charge is less stable adjacent to oxygen 21a, X = OMe) where it experiences repulsion from the two lone pairs, or at tertiary carbon (**21a** X = CH_3). Again, protonation of the pentadienyl anion ormed later on occurs on the central carbon atom.

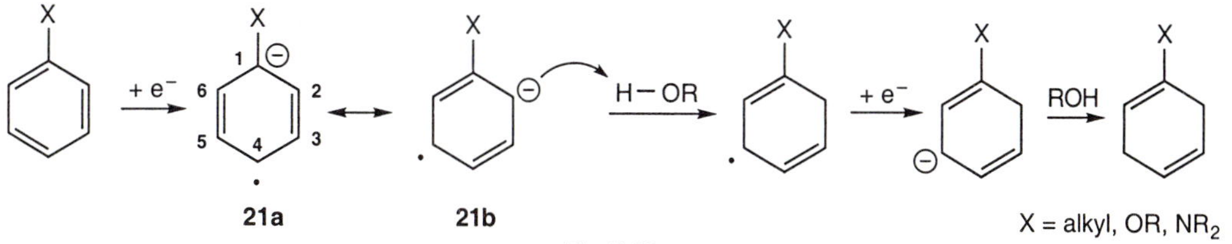

Fig. 7.16

In comparison, however, electron-deficient arenes such as benzoic acid and methyl benzoate give another substitution pattern in which the electron-withdrawing group is not in conjugation with either of the alkenes, Fig. 7.17.

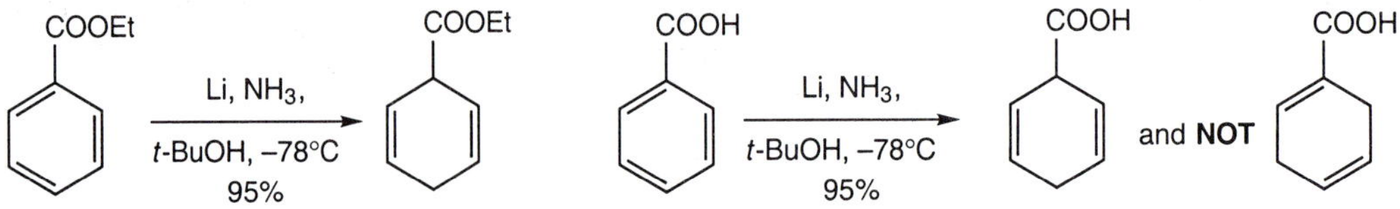

COOEt — Li, NH_3, *t*-BuOH, −78°C, 95% → COOEt

COOH — Li, NH_3, *t*-BuOH, −78°C, 95% → COOH and **NOT** COOH

Fig. 7.17

Similar mechanistic reasoning can be put to use here, when we consider that protonation of radical anion **22** is most favourable at C-4 as it leaves behind a radical that is stabilised by conjugation with the carbonyl group. Furthermore, addition of another electron forms a stable anion adjacent to the carbonyl group (i.e. an enolate), Fig. 7.18.

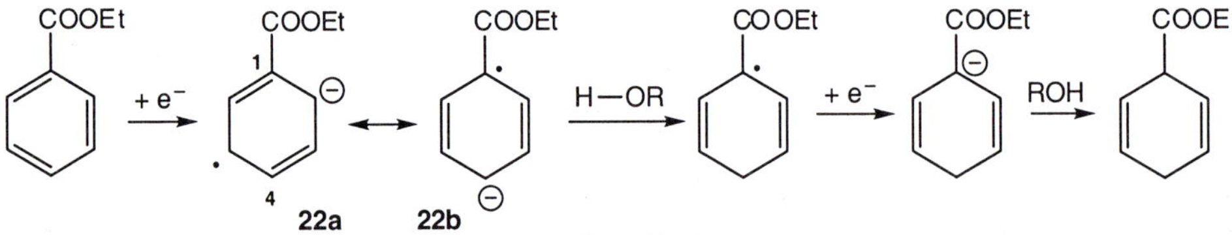

COOEt + e⁻ → COOEt (1, 4) **22a** ↔ COOEt **22b** — H—OR → COOEt — + e⁻ → COOEt — ROH → COOEt

Fig. 7.18

Remember that under Birch conditions aromatic carboxylic acids will be deprotonated. However, the carboxylate ion is still able to stabilise an adjacent negative charge.

If one limits the amount of alcohol that is added to the Birch reduction of **23** (i.e. reaction with excess sodium, ammonia and *one* equivalent of *t*-BuOH) then the enolate formed has no proton source to quench it (as the one equivalent of alcohol has been used up in protonating the radical anion). Under these conditions, the enolate can be alkylated by addition of another electrophile (E^+ is commonly an alkyl halide). This procedure is an effective way of preparing substituted cyclohexadienes, Fig. 7.19.

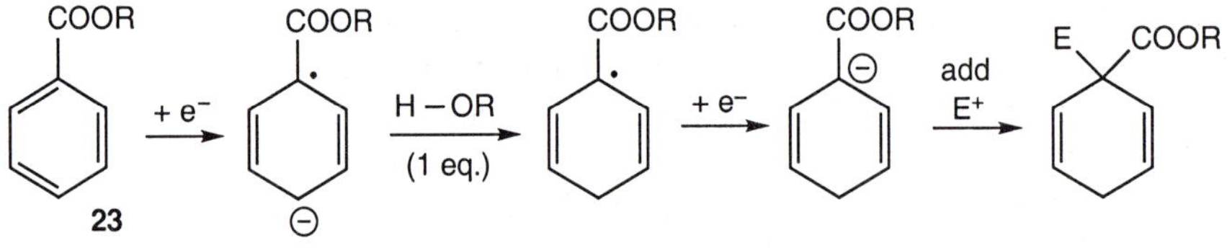

COOR **23** — + e⁻ → COOR — H−OR (1 eq.) → COOR — + e⁻ → COOR — add E^+ → E, COOR

Fig. 7.19

The two rules described above regarding orientation of the alkenes with respect to substituents can be readily applied to the reduction of more complex polysubstituted benzenes, Fig. 7.20.

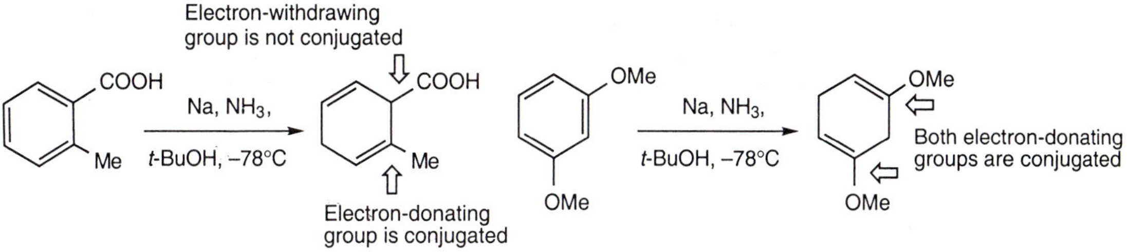

Electron-withdrawing group is not conjugated
COOH
Me
Na, NH_3, *t*-BuOH, –78°C
Electron-donating group is conjugated
OMe
OMe
Both electron-donating groups are conjugated

Fig. 7.20

Five-membered aromatic heterocycles can also be partially reduced with dium in liquid ammonia, Fig. 7.21; in each case the electron-withdrawing oup at C-2 is essential to allow efficient reduction by stabilising the radical ion formed by addition of an electron.

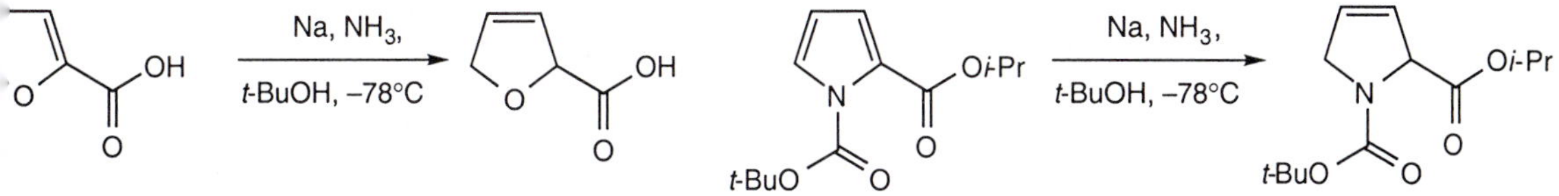

OH
O
Na, NH_3, *t*-BuOH, –78°C
O*i*-Pr
N
t-BuO

Fig. 7.21

mplete reduction of aromatic compounds

th carbocyclic and heterocyclic aromatic rings can be completely reduced th hydrogen and metal catalyst in a manner similar to that of non-aromatic enes, although conditions tend to be more forcing (one is losing maticity after all). Some transition metals are better than others for this rpose and this author recommends platinum, rhodium and ruthenium as ective catalysts.

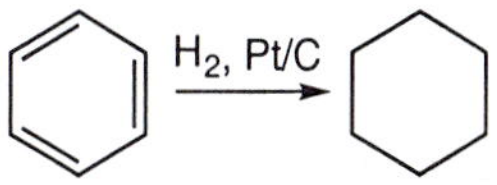

H_2, Pt/C

The mechanism of the reduction is presumed to be analogous to that lined for the reduction of alkenes and should produce cyclohexadienes and clohexenes as intermediates, although these are not isolated.

While palladium catalyst can be used for reducing aromatic compounds, it is rather inactive and so is useful for reducing alkenes in the presence of aromatic rings.

The situation becomes more interesting when substituted aromatic mpounds are reduced by hydrogenation, as stereogenic centres are formed in process. As we would expect one face of an aromatic ring to be adsorbed to the metal surface, and subsequent addition of hydrogen to come from the tal, we would predict that *cis*-diastereoisomers would be formed ferentially. This is indeed the case, although the level of stereoselectivity not always as high as that shown in Fig. 7.22.

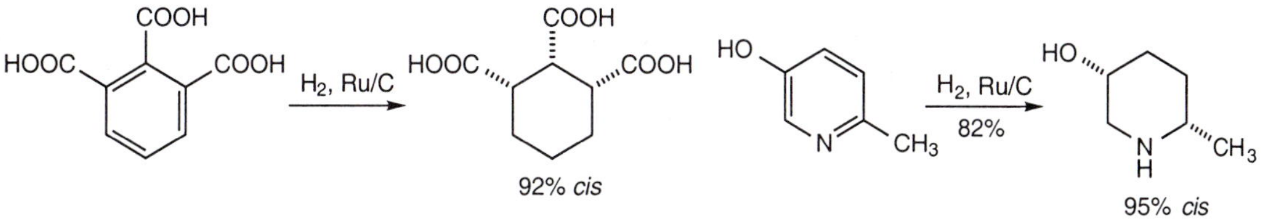

COOH
HOOC
COOH
H_2, Ru/C
92% *cis*
HO
N
CH_3
H_2, Ru/C
82%
N
H
95% *cis*

Fig. 7.22

8. Reduction of carbon–heteroatom double and triple bonds

There are many functional groups based on multiple bonds between carb and either oxygen or nitrogen. We have already dealt with the oxidati chemistry of some of these functional groups and this chapter will show y some of the corresponding reduction reactions.

8.1 Reduction of carbon–nitrogen π-bonds

Reduction of nitriles

Nitriles (or cyano groups) contain a strong triple bond between carbon a nitrogen. In terms of reactivity, nitriles are susceptible to nucleophilic atta and therefore reduction of these groups is easily effected by 'hydride' reduci agents. Let's take the partial reduction of nitriles as an example. Reaction nitriles with DIBAL-H (one equivalent) is an effective way of transformi nitriles to imines and thence to aldehydes by hydrolysis, Fig. 8.1.

DIBAL-H is di-*iso*-butylaluminium hydride

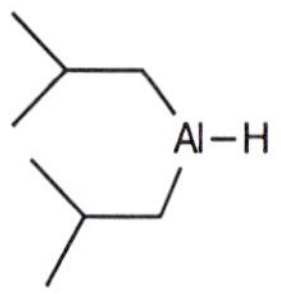

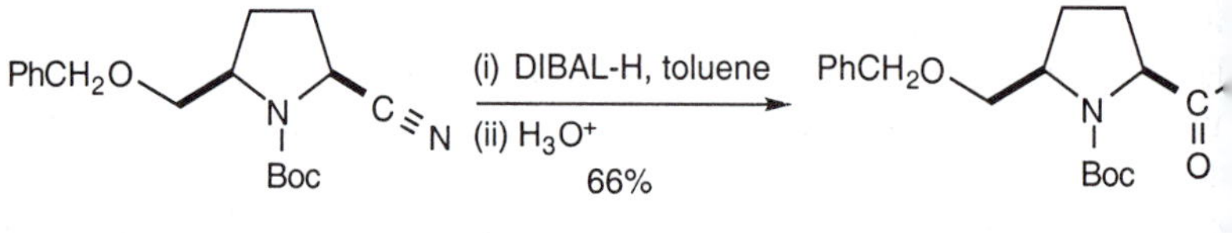

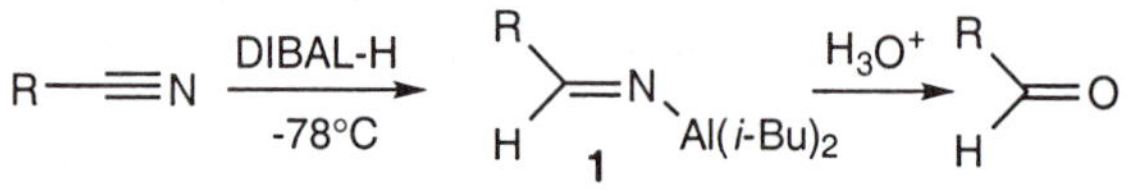

Fig. 8.1

The mechanism involves addition of the Al–H bond across the triple bc of the nitrile (this may be concerted) to form complex **1**. This complex is normally isolated but usually hydrolysed in a second step to furnish corresponding aldehyde.

Complete reduction of nitriles to amines can be achieved readily with powerful reducing agent such as $LiAlH_4$, Fig. 8.2. In the example shown t equivalents of hydride add to the triple bond forming a species that forma has two negative charges on nitrogen (of course this dianion is stabilised coordination to the Li^+ cation and the AlH_3 species formed from AlH Protonation of this dianion (this will occur on quenching the reaction duri work-up) then forms a primary amine.

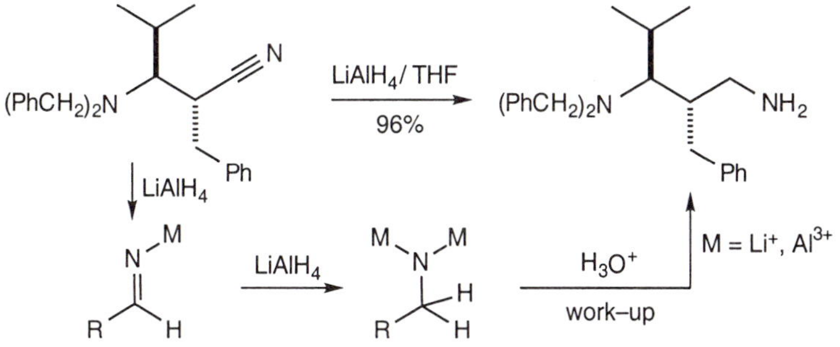

One may generally quench $LiAlH_4$ reductions with either aqueous acid or base, although the latter procedure is more practicable.

Fig. 8.2

·duction of carbon–nitrogen double bonds

duction of imines

e carbon–nitrogen double bond of an imine is readily reduced by hydride lucing agents such as sodium borohydride or lithium aluminium hydride, ;. 8.3. The example chosen shows that it is possible to reduce an imine in : presence of an ester using the weak reducing agent sodium borohydride.

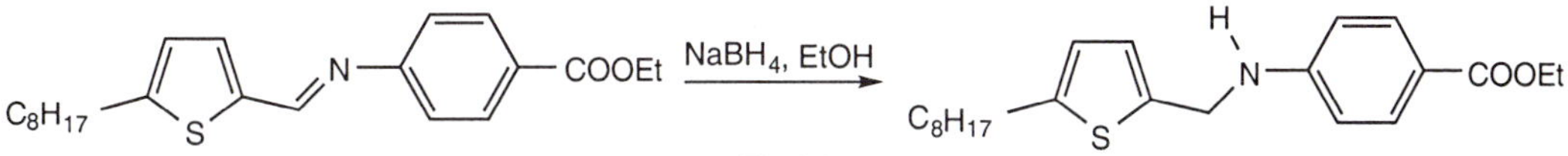

Fig. 8.3

One interesting way of increasing the reactivity of imines towards luction is to employ acid to protonate the imine nitrogen. Imines are rticularly sensitive to this type of activation because nitrogen is a basic om. This observation has given rise to a useful method of forming and lucing imines *in situ*; the overall transformation of carbonyl group to ine and then to amine is known as reductive amination, Fig. 8.4. The ndensation with acetone is a particularly useful method of putting an propyl group on nitrogen: certainly much better than (slow) S_N2 reaction an amine and *iso*propyl bromide which will have problems of elimination d over-alkylation.

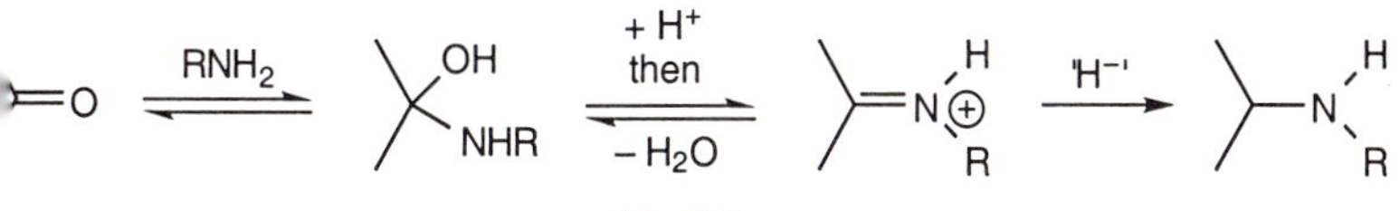

Fig. 8.4

The ketone is not reduced at an appreciable rate under these conditions because the pH is not low enough to protonate it.

Clearly, if this method is going to work then we require a reducing agent rmally a source of H^-) that will not react as a base in the presence of an dic medium. Sodium cyanoborohydride ($NaBH_3CN$) is a deactivated rohydride reagent that fits the bill nicely because it is compatible with lutions at pH 3–6 and is also sufficiently unreactive to avoid reducing the bonyl group. Thus, primary amines can be converted to secondary, and condary amines to tertiary, Fig. 8.5.

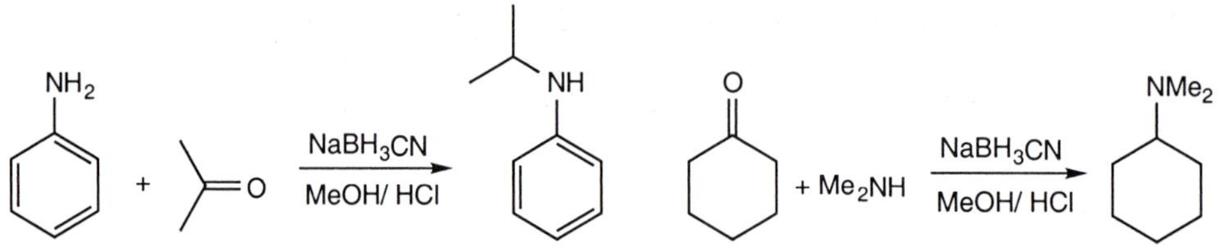

Fig. 8.5

On occasion it can be difficult to control the reaction of primary amines that they stop at the secondary amine stage. The main difficulty stems fro the ability of the secondary amine product to react further *via* an iminium i (this is positively charged and reactive), Fig. 8.6. In fact this facet of ami chemistry can be put to good use in a *double* reductive alkylation where primary amines are converted directly to tertiary amines: both alkyl grou that are introduced have to be the same and an excess of carbonyl compou will be required. Reductive methylation, using formaldehyde, is probably t most common use of this reaction, Fig. 8.6. Of course, it is not possible over-alkylate the amine and form quaternary ammonium salts as the tertia amine products (e.g. **2**) cannot condense with another mole of formaldehyd

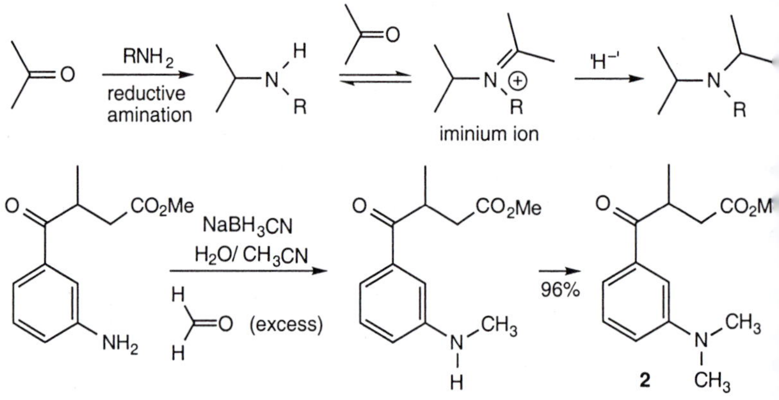

Fig. 8.6

The Eschweiler–Clark reaction is related to the reductive methylati reaction and also uses formaldehyde to form iminium ions, Fig. 8.7.

Monomeric formaldehyde (CH_2O) is unstable in solution and the polymer $(CH_2O)_n$ is used instead.

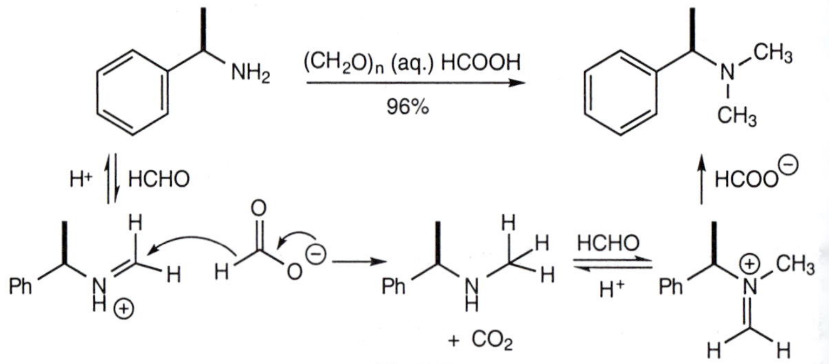

Fig. 8.7

The main point of interest is the use of formic acid (HCOOH) as both a proton *and* a hydride source, Fig. 8.7. The key step is transfer of hydride ion from the formate anion onto an iminium ion electrophile, thus liberating carbon dioxide.

Partial reduction of pyridines

We have already discussed the complete reduction of pyridines in Chapter 7. It is, however, possible to reduce partially the pyridine nucleus (and other related heterocycles) with hydride reducing agents. Pyridine itself does not react with $NaBH_4$ unless electron-withdrawing groups are attached to activate the heterocycle to nucleophilic attack. So, reduction of **3**, Fig. 8.8 with $NaBH_4$ in ethanol gives the tetrahydropyridine **4**; a plausible mechanism is shown which involves the intermediacy of a 1,4-dihydropyridine (in the mechanism shown, ethanol can act as the H^+ source).

You should remember the 1,4-dihydropyridine intermediate shown in Fig. 8.8 as it is closely related to a reducing agent used by nature that we shall come across later on.

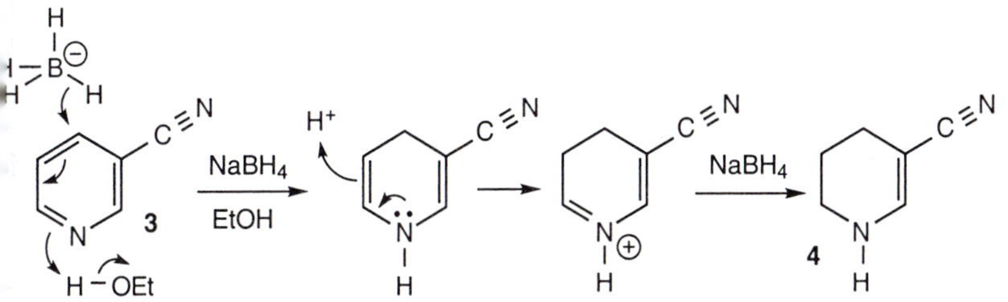

Fig. 8.8

Placing a positive charge on the nitrogen of a pyridine (i.e. making a pyridinum salt) is an obvious way of increasing the electrophilicity of this heterocycle. Accordingly, Wenkert showed that the pyridinium salts reduce twice with sodium borohydride, Fig. 8.9.

Reduction of *bis*-pyridinium salt **5** was a key step in a recent synthesis of the alkaloid keramaphidin B. This reaction also illustrates the fact that isolated alkenes are not attacked by sodium borohydride.

Fig. 8.9

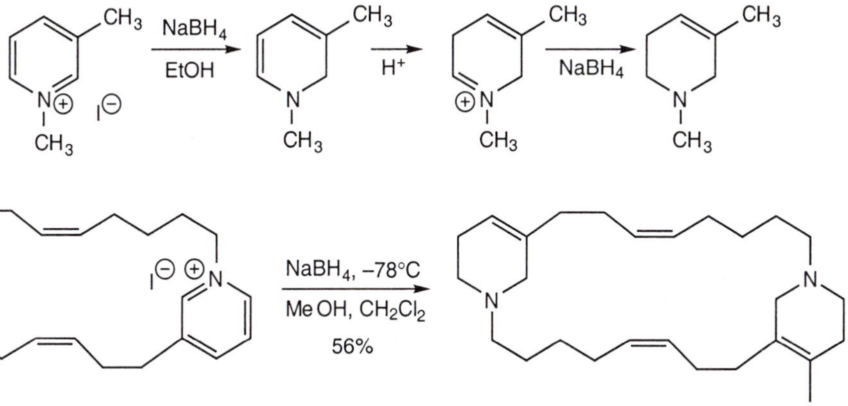

8.2 Reduction of carbon–oxygen π-bonds

Reduction of aldehydes and ketones

The reduction of aldehydes to primary alcohols and ketones to secondary alcohols is normally easy to accomplish using sodium borohydride or lithium aluminium hydride. Let's consider $NaBH_4$ first as this mild reducing agent can react with aldehydes and ketones without reducing carboxylic acids, esters or amides, Fig. 8.10. The mechanism of reduction certainly involves addition of hydride from BH_4^- to the carbonyl group and the developing charge on the oxygen is quenched by reaction with the alcohol solvent.

It is noteworthy that sodium borohydride reacts with alcohols (e.g. MeOH) to give alkoxyborohydrides of varying composition. Each of these species is able to donate hydride to a carbonyl group and generally the more oxygen ligands on boron then the more reactive the reducing agent.

$NaBH_4$ + MeOH → $NaBH_3OMe$ + H_2 —MeOH→ $NaBH_2OMe_2$ + H_2

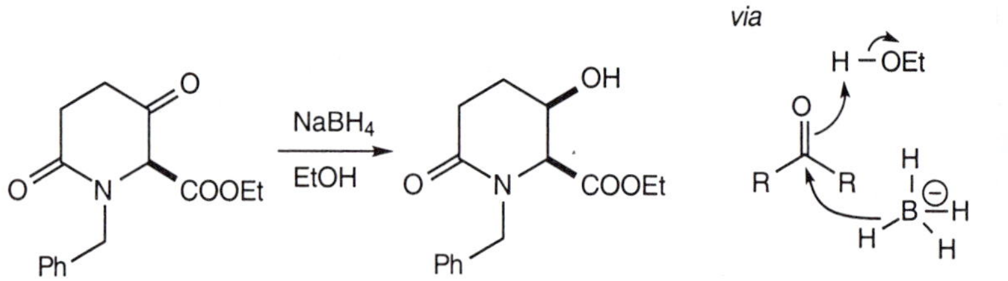

Fig. 8.10

Lithium aluminium hydride acts in the same manner as sodium borohydride but is many times more reactive. The mechanism by which $LiAlH_4$ reduces carbonyls is different to that shown above for borohydride. First, activation of the carbonyl with the Li^+ cation is essential to initiate reduction. Addition of hydride to the activated carbonyl ensues and the negatively charged oxygen that results from this process is stabilised by coordination to both Li^+ and AlH_3, Fig. 8.11.

The importance of Lewis acid activation of the carbonyl is underscored by the observation that in the presence of a crown ether that binds Li^+, reduction of ketones with $LiAlH_4$ is very slow.

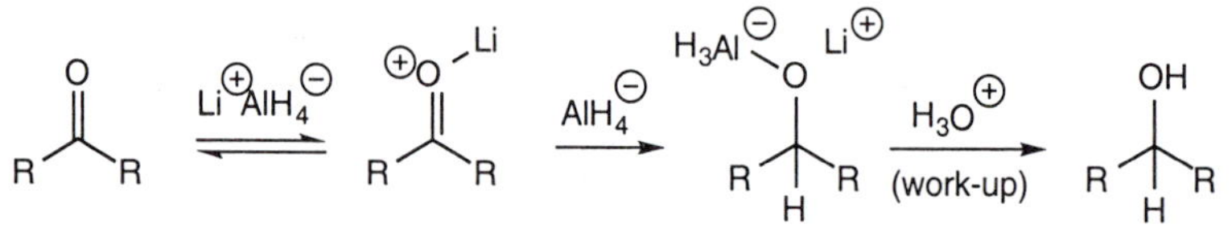

Fig. 8.11

A fundamentally different way of reducing aldehydes and ketones utilises hydride transfer from a secondary alcohol. The Meerwein-Pondorf-Verley reduction involves reaction of a carbonyl with Al(O*i*-Pr)$_3$ in *i*-PrOH solvent, Fig. 8.12.

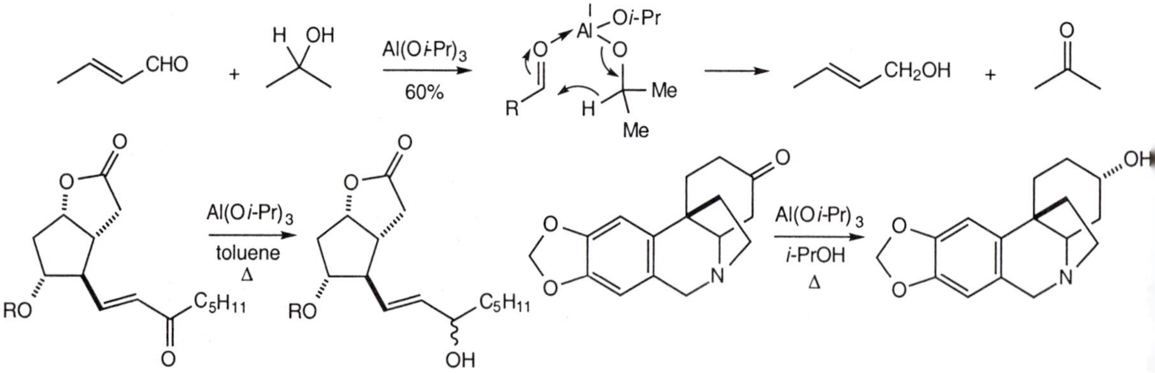

Fig. 8.12

The reaction is presumed to proceed *via* a six-membered cyclic transition state with internal hydride transfer. Moreover, this reaction is an equilibrium which is shifted to the product side by using excess *i*-PrOH or by removing the volatile acetone by-product. Aluminium alkoxides are good reagents for the Meerwein-Pondorf-Verley reduction because they are only weakly basic and so avoid complications arising from the condensation reactions of aldehyde or ketone enolates.

The equilibrium established in the Meerwein-Pondorf-Verley reduction may be reversed by using $Al(O\textit{i}\text{-}Pr)_3$ and excess acetone. The result is a method of oxidising primary and secondary alcohols known as the Oppenauer oxidation.

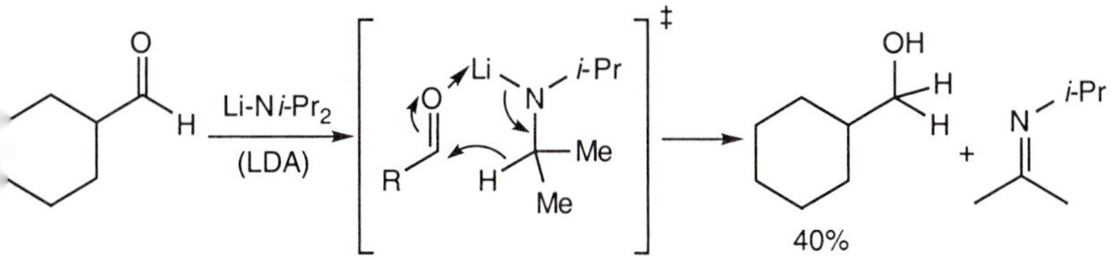

A related hydride transfer from lithium amides can also be observed: a common by-product from the attempted enolisation of aldehydes by LDA is the primary alcohol. It appears that hydride transfer from the amide to the electrophilic carbonyl group is responsible, Fig. 8.13.

Fig. 8.13

Selectivity in the reduction of aldehydes, ketones and enones

Reduction of α,β-unsaturated ketones can lead to problems derived from a lack of regioselectivity during hydride attack. For example, reduction of cyclopentenone by sodium borohydride can lead to two products, **6** and **7**, Fig. 8.14. The allylic alcohol **6** derives from 1,2-attack of hydride onto the ketone, while the fully reduced compound originates from 1,4-attack, tautomerism of the enol and further reduction of the ketone product. Clearly, some means of controlling this regioselectivity is required if the reaction is to be used successfully. To this end, an important modification of sodium borohydride was made by Luche who added $CeCl_3$ to a methanolic solution and observed complete 1,2-selectivity: so it is possible to reduce the ketone but not the conjugated alkene. Remember the opposite regioselectivity is possible, i.e. that sodium in liquid ammonia reduces the alkene selectively, rather than the ketone (see Chapter 7) .

Luche showed that Ce^+ promotes the reaction between methanol and $NaBH_4$ forming alkoxyborohydrides *in situ*. These reducing agents may themselves react with a selectivity that is different from that of the parent $NaBH_4$.

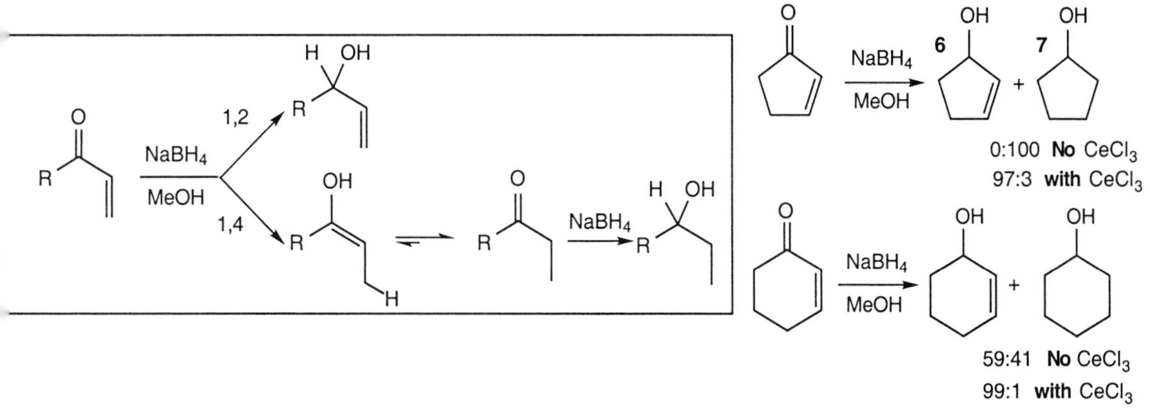

Fig. 8.14

The combination of cerium trichloride and sodium borohydride is not so selective for the reduction of ketones in the presence of conjugated aldehydes as these aldehydes are less willing to form hydrates *in situ*.

Another useful reaction which involves the combination of $CeCl_3$ and sodium borohydride is the selective reduction of ketones in the presence of aldehydes, Fig. 8.15. At first this seems counterintuitive as aldehydes are more reactive electrophiles than ketones and, if anything, it should be possible to reduce aldehydes rather than ketones: so, how is this selectivity achieved? The answer does indeed lie in the increased reactivity of aldehydes relative to ketones and has its origins in the formation of hydrates from the aldehyde (facilitated by cerium trichloride) and not from the ketone. This means that the aldehyde is 'protected' against reduction *in situ* and so only the ketone can react with borohydride in the normal manner: presumably the aldehyde is regenerated from the hydrate during aqueous work-up.

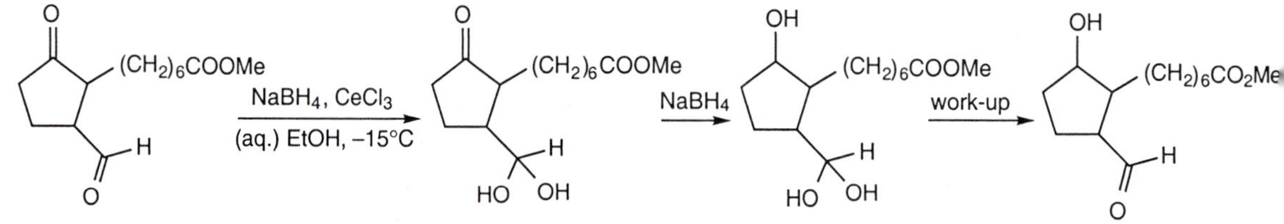

Fig. 8.15

Stereoselectivity during ketone reduction

Reduction of a ketone within a chiral compound normally gives rise to diastereoisomers: steric effects can dominate reductions with the hydride reducing agent approaching a carbonyl from the least hindered face, Fig. 8.16. Other factors such as chelation between the substrate and the reducing agent can also influence the facial selectivity upon reduction.

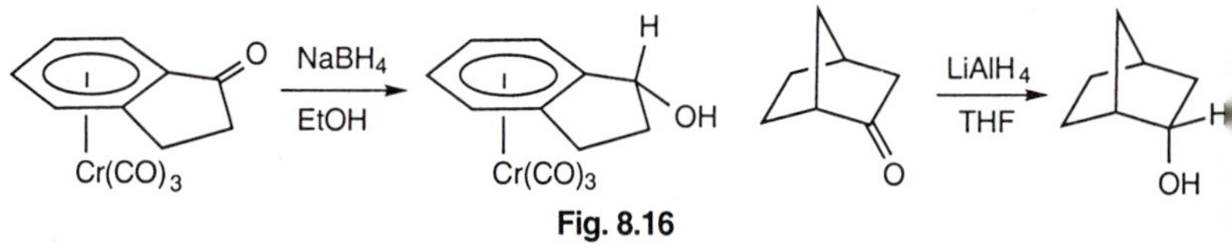

Fig. 8.16

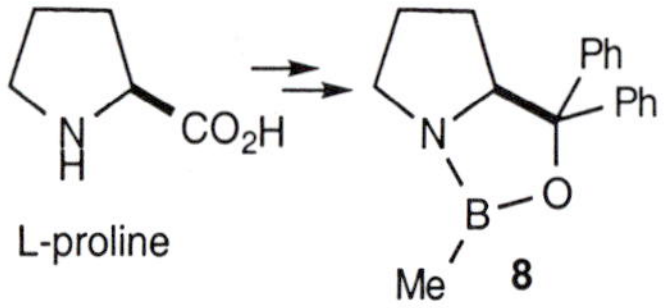

L-proline

The oxazaborolidine complex **8** is both Lewis acidic (at boron) and Lewis basic (at nitrogen). This feature of the catalyst, which is essential to its activity, is exaggerated by the cyclic nature of the ring system which is too constrained to allow ideal overlap between the nitrogen lone pair and a p orbital on boron.

In recent years there have been advances in the use of enantiomerically pure reducing agents to impose stereoselectivity during the reduction of ketones. Under normal circumstances, reduction of a prochiral ketone leads to a racemic mixture of secondary alcohols. However, modification of the reducing agent with a chiral (non-racemic) ligand changes this outcome and the production of secondary alcohol as single enantiomers is now commonplace, Fig. 8.17. One of the more useful chiral reducing agents is the CBS oxazaborolidine devised by Corey; the chiral ligand **8** derives from proline. Compound **8** has the ability to accelerate the reduction of ketones by diborane and one of the advantages of this reagent is that it may be used as a catalyst with only very small amounts required for a reduction. The source of 'hydride' for the reduction comes from diborane or catecholborane, which must be used in stoichiometric amounts.

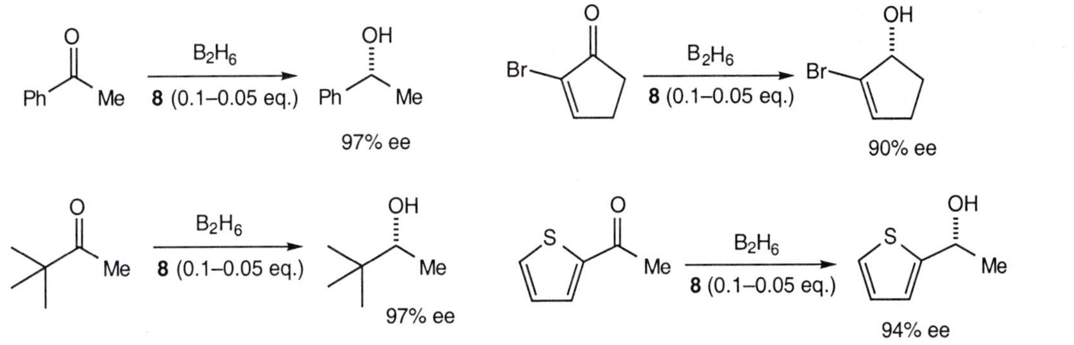

Fig. 8.17

The chiral catalyst **8** exerts its influence by binding to both borane and the ketone substrate, Fig. 8.18. In binding both partners the catalyst activates the ketone and borane to reduction and this follows *via* a six-membered transition state. The formation of a single enantiomer of alcohol product has its origins in the ordered binding between Lewis acidic **8** and the ketone; this is most favourable on the oxygen lone pair opposite the larger of the two groups attached to the ketone. Moreover, both reagents bind to the relatively unhindered *exo* face of the bicyclo ring of the catalyst, so creating an environment for selectivity during the hydride transfer step.

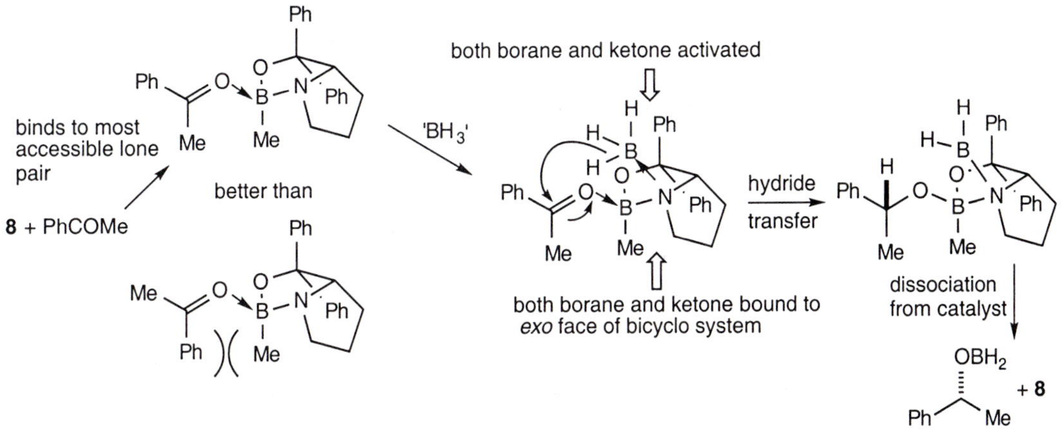

Fig. 8.18

After reduction is complete, the alkoxide and BH_2 dissociate from the catalyst to form a boronate (e.g. $ROBH_2$) and regenerate **8**. As we have seen before, the boronate is hydrolysed to an alcohol upon work-up.

Reduction of carbonyl groups in nature

Nature routinely reduces carbonyl groups during the biosynthesis of all sorts of natural products (fatty acids and carbohydrates are just two examples from many). The enzymes that are used for this process require a co-factor called NADPH, Fig. 8.19. Although this is a complicated looking structure it is only the dihydropyridine portion of the molecule that need concern us.

The oxidised form of NADPH, called NADP$^+$, is used by nature as an oxidising agent in what is essentially the reverse of the reaction shown in Fig. 8.19.

Interestingly, it has been shown that with some enzymes the H_R hydride is transferred and with others it is H_S. This selectivity appears to be correlated to the reactivity of the carbonyl substrate.

The mechanism by which NADPH reduces carbonyl compounds is similar to that shown for sodium borohydride, with the dihydropyridine ring acting as a source of hydride. It has been shown that only one of the two hydrogen atoms at the C-4 position of the dihydropyridine ring is removed during reduction and that reduction of ketones (even unnatural ones) frequently gives products with high enantiomeric excess. This degree of control need not be surprising in the presence of enzymes which are capable of controlling stereochemistry with consummate ease.

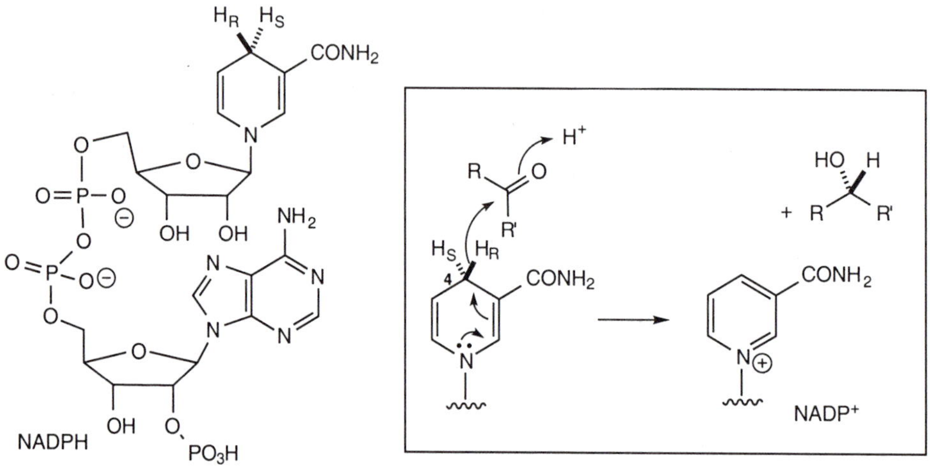

NADPH

NADP$^+$

Fig. 8.19

Reduction of esters

The reduction of esters to primary alcohols is a straightforward task that is best accomplished by a powerful reducing agent such as $LiAlH_4$ (alternatives include *i*-Bu_2AlH and $LiBH_4$); $NaBH_4$ is not reactive enough to accomplish the reduction of simple esters. Under such reaction conditions, hydride addition to the carbonyl gives a tetrahedral intermediate (stabilised by coordination to aluminium and lithium) that collapses to form an aldehyde and (as aldehydes are more reactive than esters) this is quickly reduced to the corresponding alcohol, Fig. 8.20. It is interesting to note that the aluminate species formed *in situ* by replacement of hydride ligands by oxygen ligands (e.g. $R'OAlH_3^-$) are also active reducing agents, but in this case they are *less* potent than the parent $LiAlH_4$.

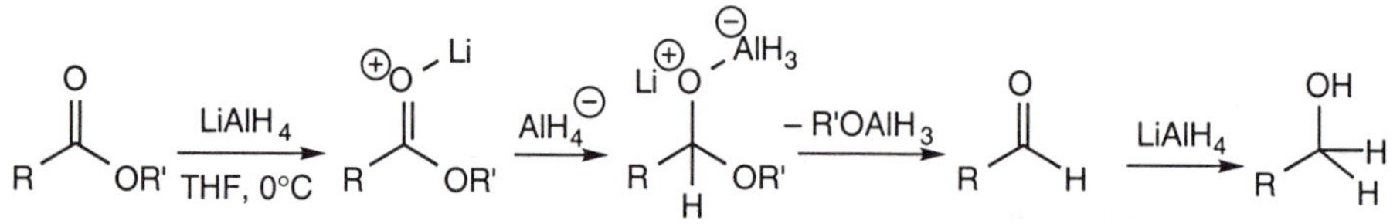

Fig. 8.20

$LiBH_4$ reduces esters while $NaBH_4$ does not. The difference lies in the ability of Li^+ to activate the ester carbonyl to reduction. Na^+ is insufficiently Lewis acidic to do this.

Remember that although reduction of RCOOR' may be considered as a means of preparing RCH_2OH, the alcohol product R'OH may also be a desirable material.

Partial reduction of esters and lactones

Esters may be partially reduced to aldehydes if one equivalent of a hydride nucleophile is added at low temperature. In practice, DIBAL-H (1 eq.) in toluene is most often used for such selective reductions. It is important that the tetrahedral intermediate formed upon addition of 'H^-' does not collapse *in situ* or else the resulting aldehyde will be reduced again; this can be rather difficult to control. Of course, on work-up, treatment with a proton source and warming, the tetrahedral intermediates derived from these compounds will collapse to form an aldehyde. So, reduction of esters **9** and **10** with DIBAL-H (or the isotopically labelled DIBAL-D) gave two aldehydes in reasonable to excellent yield, Fig. 8.21.

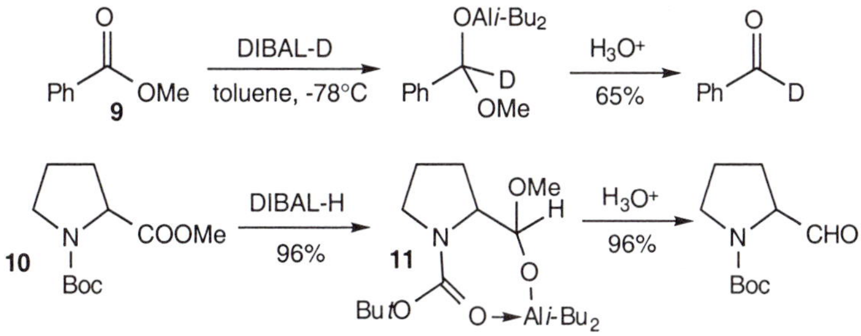

Factors which stabilise the tetrahedral intermediate help to prevent its collapse and further reduction; so, the tetrahedral intermediate **11** is expected to be especially stable due to chelation between the alkoxide counter-ion and the *N*-Boc group.

Fig. 8.21

Another related reaction is the partial reduction of lactones to cyclic hemiacetals, known as lactols. This reaction is also best accomplished with DIBAL-H at low temperature, Fig. 8.22. During the reduction of lactones, collapse of the tetrahedral intermediate would expel an alkoxide in an intramolecular sense. Clearly there is not as much entropic advantage in opening a ring as there would be in expelling a separate molecule of alkoxide and so the partial reduction of lactones is usually an efficient process.

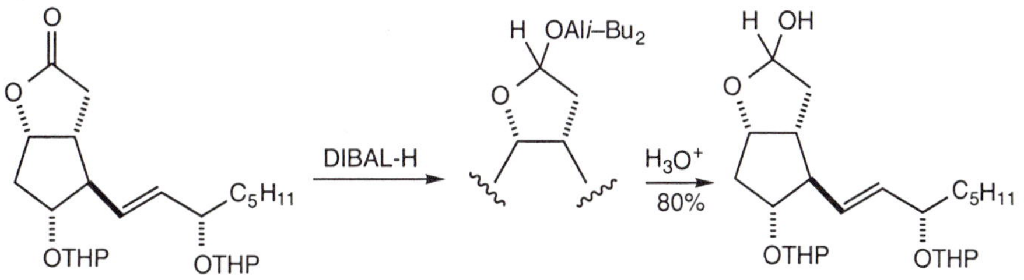

Fig. 8.22

Reduction of carboxylic acids

In practice, carboxylic acids are more difficult to reduce than esters using hydride reducing agents, because they form inert carboxylate salts with concomitant evolution of hydrogen. Lithium aluminium hydride can be used to reduce acids but this can require vigorous conditions and clearly other reducible functional groups are unlikely to survive such conditions intact, Fig. 8.23.

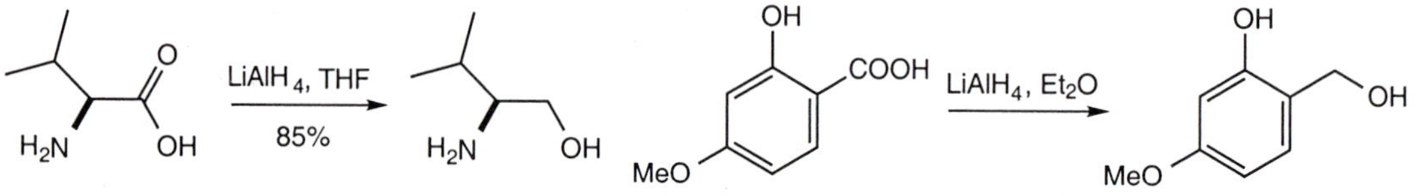

Fig. 8.23

A more convenient way of reducing acids to primary alcohols involves prior activation of the acid as a mixed anhydride or acid chloride followed by reduction of this reactive intermediate with sodium borohydride, Fig. 8.24.

Both anhydrides and acid chlorides are particularly electrophilic and so this two-step procedure is an excellent way of reducing acids in the presence of esters. The mixed anyhdride shown in Fig. 8.24 is most reactive at the C=O that originally belonged to the carboxylic acid: the other carbonyl is not as electrophilic because of overlap of the C=O π– system with electrons on two oxygen atoms.

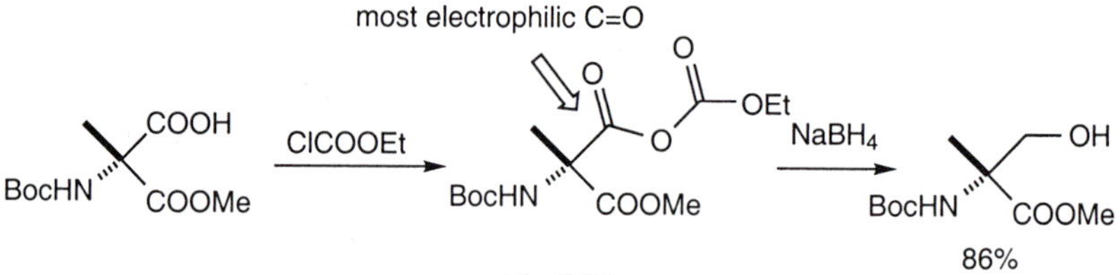

Fig. 8.24

Diborane (B_2H_6) is a useful reagent for the reduction of acids to primary alcohols. However, unlike hydride donors such as $LiAlH_4$, this reagent reacts faster with acids than it does with esters, Fig. 8.25.

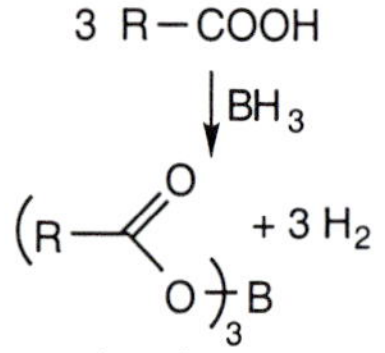

Fig. 8.25

acyloxy borane

The initial reaction of a carboxylic acid with borane evolves hydrogen to produce acyloxy boranes. Unlike the carboxylate salts mentioned earlier, such boron complexes are particularly reactive towards excess reducing agent and reduction to the aldehyde and then alcohol ensues. The reason for the enhanced reactivity of acyloxy boranes, relative to esters and acids, may have its origins in an overlap of the oxygen lone pair with the empty p-orbital on boron. This effect should in turn reduce the effectiveness of the overlap between the oxygen lone pair and the adjacent π-system of the C=O bond.

The enhanced reactivity of acyloxy boranes is illustrated by the fact that they can be reduced to an alcohol by $NaBH_4$.

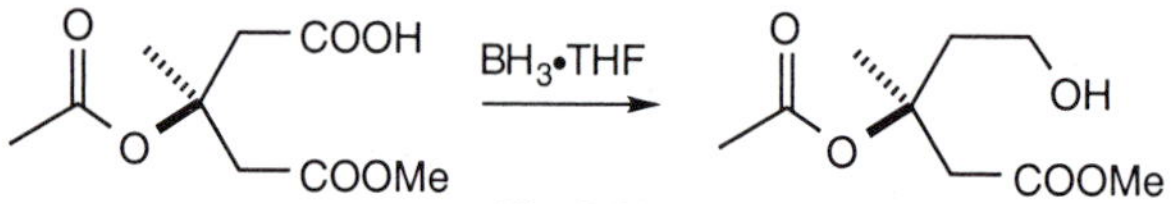

The reduction of amides

The reduction of amides with strong hydride donors such as $LiAlH_4$ does not follow the same pattern as that observed for esters and acids. Instead of an alcohol being formed, the product is an amine: i.e. $LiAlH_4$ is a means of deoxygenating amides, Fig. 8.26. Several groups are tolerated in this reduction (as long as they do not react with $LiAlH_4$ themselves) and both acyclic and cyclic amides (lactams) are good substrates.

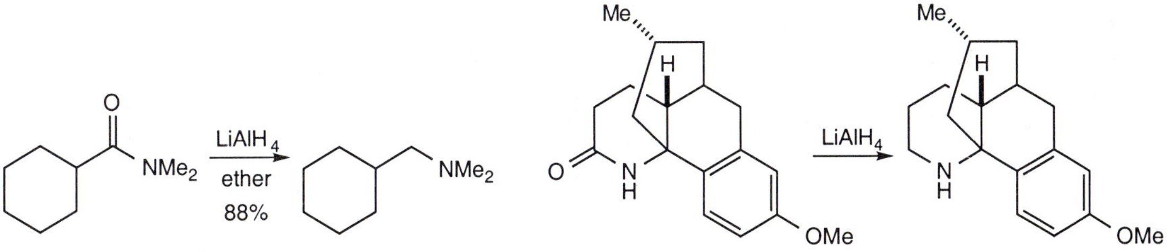

Fig. 8.26

The mechanism of this reduction involves coordination of a lithium cation to the amide carbonyl, thus activating it to nucleophilic attack by AlH_4^-. The resulting tetrahedral intermediate can now expel the oxygen atom (rather than the nitrogen atom which is a poor leaving group) with the incipient negative charge being stabilised by the oxophilic Li and Al. The iminium ion that results from this process will be reduced rapidly by $LiAlH_4$ thus forming an amine, Fig. 8.27.

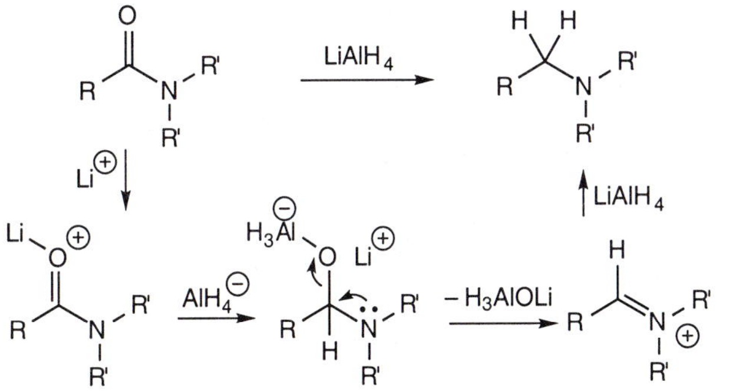

A useful way of making substituted amines involves the reaction of an amine with an acid chloride, followed by reduction of the amide with $LiAlH_4$.

Fig. 8.27

Borane is also a very good reagent for deoxygenating amides to furnish amines. However, borane is a strongly Lewis acidic reducing agent that reacts fastest with nucleophilic carbonyl groups. So, it should come as no surprise to find that amides can be reduced to amines in the presence of the ostensibly more reactive ester group. Other selective reductions are possible with this reagent, Fig. 8.28.

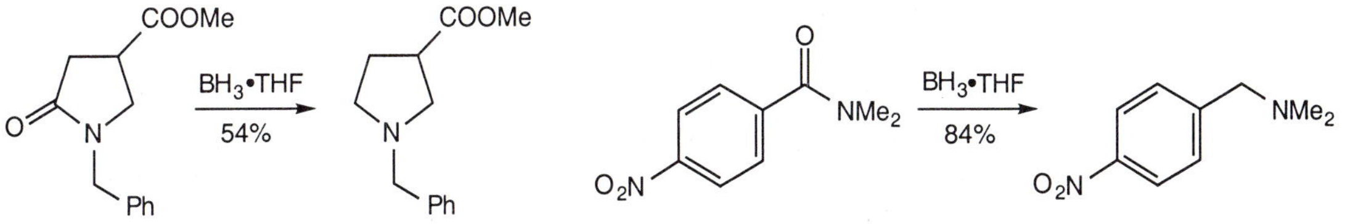

Fig. 8.28

Partial reduction of amides

Amides can be partially reduced to aldehydes if the tetrahedral intermediate formed after addition of hydride can be stabilised (this is analogous to the partial reduction of esters). In this scenario, work-up with a proton source yields a hemiaminal which hydrolyses to form an aldehyde. However, there is not a general-purpose reducing agent capable of achieving the transformation of amides to aldehydes. Perhaps the best reagent available is the ATE complex **12**, derived from the addition of BuLi to DIBAL-H, Fig. 8.29. Sodium in liquid ammonia also has some use in this reaction. However, neither reagent is useful for the reduction of primary or secondary amides.

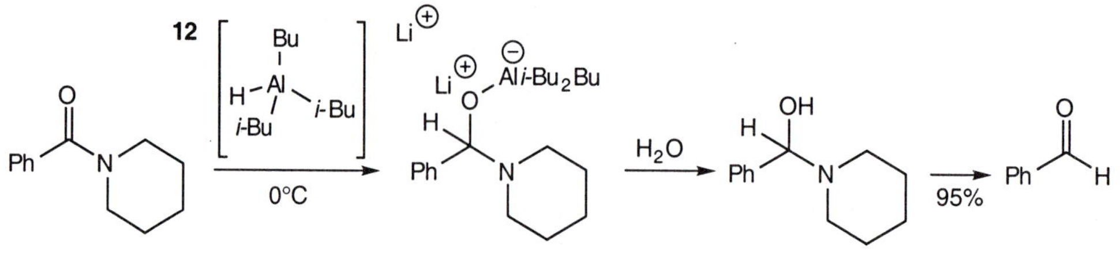

12 [Bu H–Al i-Bu i-Bu] $Li^{\oplus}$, 0°C

Ph, O, N — $Li^{\oplus}$ O–$^{\ominus}$Al*i*-Bu_2Bu, H, Ph, N — H_2O — OH, H, Ph, N — 95% — Ph, O, H

Fig. 8.29

Weinreb's amides may be prepared from the corresponding acid chloride and NHMeOMe or by reaction of an ester with Me_3Al and NHMeOMe.

One good way of reducing amides selectively involves the formation of Weinreb amides, **13**, Fig. 8.30. These derivatives react with either $LiAlH_4$ or DIBAL-H at low temperature and form stable, chelated intermediates. So, on work-up, the corresponding aldehydes are formed in high yields.

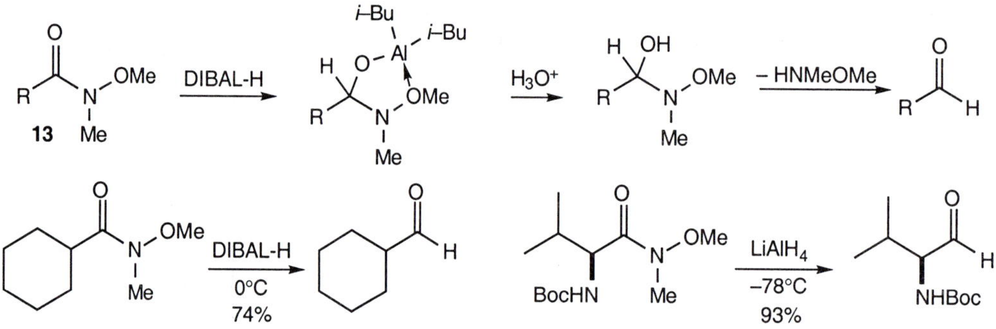

O, R, N, OMe, Me, 13 — DIBAL-H — *i*-Bu, *i*-Bu, Al, H, O, OMe, R, N, Me — H_3O^+ — H, OH, R, N, OMe, Me — – HNMeOMe — O, R, H

O, N, OMe, Me — DIBAL-H, 0°C, 74% — O, H

O, N, OMe, Me, BocHN — $LiAlH_4$, –78°C, 93% — O, H, NHBoc

Fig. 8.30

9. Reductive cleavage reactions

In this final chapter we will look at reactions which completely cleave the bonds between carbon and electronegative elements, replacing them with bonds to hydrogen. Here you will find examples of reductions which cleave purely σ-bonds and also reactions which cleave both σ- and π-bonds.

9.1 Reduction of carbon–nitrogen σ-bonds

Reductive cleavage of amines and amides

Amines that are substituted with a benzyl group (Bn or $PhCH_2$) may be broken by reduction with hydrogen and a transition metal catalyst, Fig. 9.1 (the benzylic carbon–nitrogen bonds that are cleaved are shown with a dotted line). In practice, this property of the benzyl group means that it is often used as a protecting group for amines and occasionally amides. Generally, tertiary amines are easier to deprotect this way than secondary, which in turn are more reactive than primary amines. See Fig. 9.5 for an outline of the mechanism for removal of a benzyl group.

Reductive cleavage using hydrogen and a catalyst is known as hydrogenolysis (cf. hydrogenation).

Most hydrogenolysis reactions are accelerated by the addition of acid (acetic acid is a good one) to the reaction mixture.

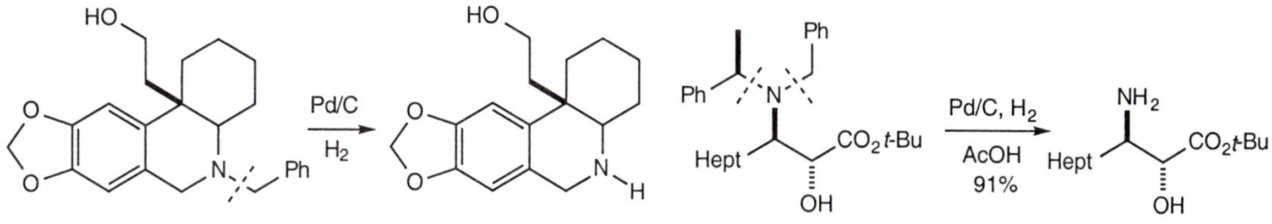
HO
O
O
N
Ph
Pd/C
H2
HO
O
O
N
H
Ph
Ph
N
Hept
CO2t-Bu
OH
Pd/C, H2
AcOH
91%
NH2
Hept
CO2t-Bu
OH

Fig. 9.1

Sodium in liquid ammonia (with a proton source such as *t*-butanol added) is a good alternative to hydrogenolysis for deprotecting benzyl amines and amides, Fig. 9.2.

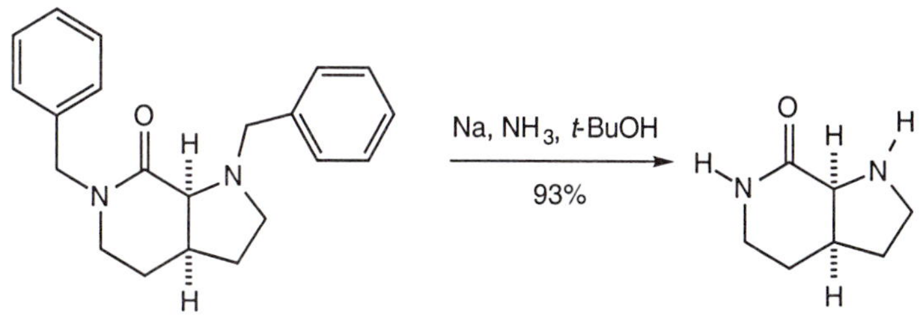
O
H
N
N
H
Na, NH3, t-BuOH
93%
O
H
H
H
N
N
H

Fig. 9.2

Presumably, the mechanism of this reaction involves addition of an electron to the aromatic π-system (just like the Birch reduction). In this case, the intermediates (e.g. **1** and **2**, Fig. 9.3) formed during the reduction can expel a leaving group from the benzylic position (X^- in Fig. 9.3). As you will see later on, there are other heteroatoms that can be protected as their benzyl derivatives (this is simply a change in the identity of X) and sodium in ammonia is a good method of deprotection for most of them.

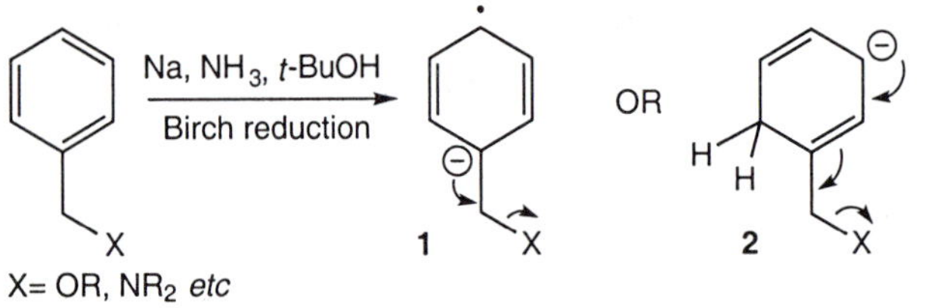

Na, NH_3, *t*-BuOH
Birch reduction
X= OR, NR_2 *etc*
OR

Fig. 9.3

9.2 Reductive cleavage of carbon–oxygen bonds

Reductive cleavage of benzyl ethers, esters and carbamates

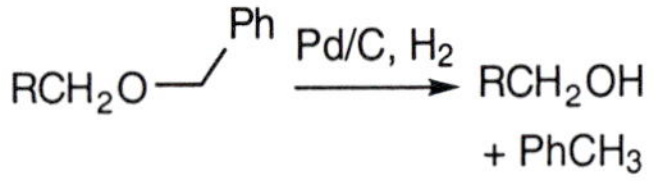

In common with the nitrogen functionality discussed earlier, benzyl derivatives are often used as protecting groups for oxygen-based functional groups, most notably protection of alcohols as benzyl ethers and carboxylic acids as benzyl esters. Removal of these groups may be achieved by reduction with hydrogen and a transition metal catalyst (palladium on charcoal is probably the most widely used). Figure 9.4 shows the cleavage of a benzyl ester and the double hydrogenolysis of a benzylidene acetal.

RCH_2O–CH_2Ph $\xrightarrow{Pd/C,\ H_2}$ RCH_2OH + $PhCH_3$

Normally, oxygen-protecting groups of this kind are easier to deprotect than the corresponding nitrogen-protecting groups. Of course, isolated alkenes are normally hydrogenated during hydrogenolysis and this process can be avoided only if the alkene is highly substituted and approach to the catalyst sterically hindered.

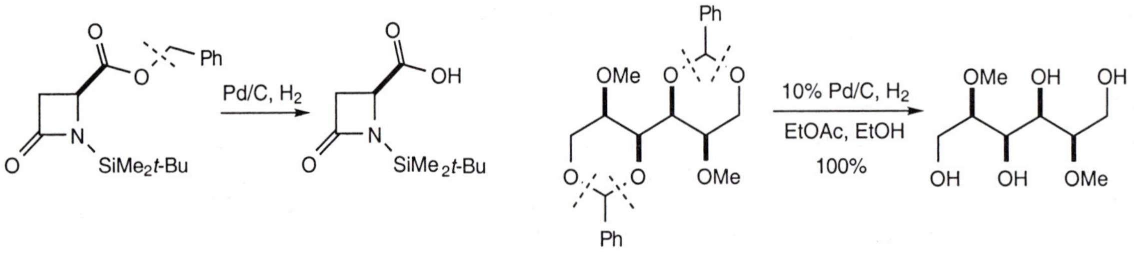

Pd/C, H_2
$SiMe_2t$-Bu
10% Pd/C, H_2
EtOAc, EtOH
100%

Fig. 9.4

The mechanism of hydrogenolysis of benzyl derivatives is complex and Fig. 9.5 is intended to provide an outline of one plausible mechanism.

Hydrogenolysis is preceded by adsorption of the aromatic group onto the ırface of the catalyst (of course, adsorbed hydrogen atoms are also present on ıis surface). Bond breaking may then occur *via* nucleophilic attack of a ıemisorbed hydrogen in a manner reminiscent of an S_N2 reaction: this ıechanism would explain why acid promotes hydrogenolysis (as it makes the ɜteroatom into a better leaving group) and also why some hydrogenolysis actions proceed with inversion of configuration.

Studies on the hydrogenolysis of **3** with palladium showed almost complete inversion during C–O bond cleavage.

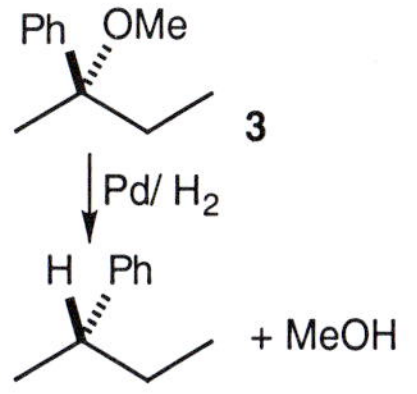

Unlike hydrogenation, homogeneous transition metal catalysts do not ɛem to promote hydrogenolysis which implies that more than one metal om may be required for the hydrogenolysis process.

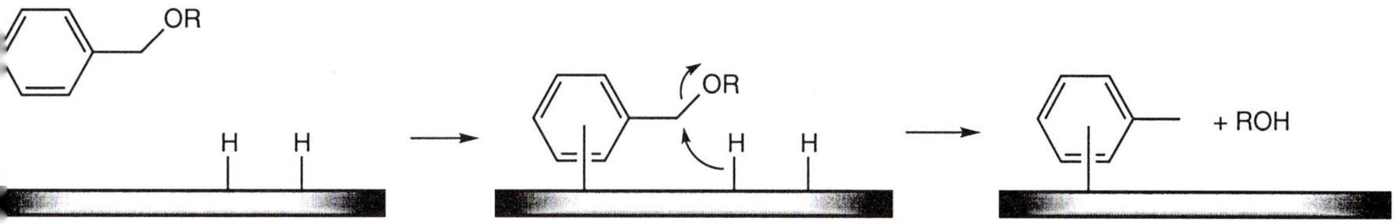

Fig. 9.5

Recent reports have indicated that the naphthyl analogue of the benzyl oup (NAP) is rather easy to hydrogenolyse and this may have some use in nthesis where there is always a need for selectivity during removal of otecting groups. Presumably the relative ease of hydrogenolysis of the AP group, relative to the benzyl group, has its origins in the high affinity the flat extended π-surface for the metal catalyst.

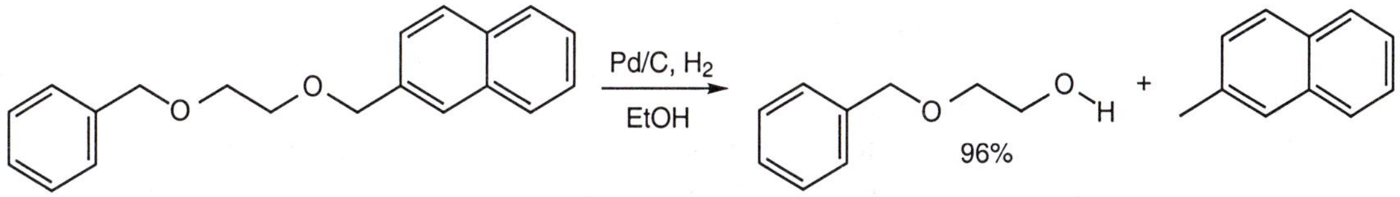

Fig. 9.6

Sodium in liquid ammonia is another useful method of deprotecting benzyl oups (see Fig. 9.3, X = OR).

leavage of carbonylbenzyloxy (Cbz) groups

he carbonylbenzyloxy group (Cbz or sometimes just called Z after its ventor Zervas) is a useful protecting group for amines (e.g. $_2$NCOOCH$_2$Ph = R$_2$NCbz = R$_2$NZ). This group is stable to hydrolysis and ost oxidising and reducing agents: amines are normally protected by action with CbzCl (i.e. ClCOOCH$_2$Ph). Deprotection can be effected by ydrogenolysis and it is rupture of a carbon–oxygen bond that leads to moval of the Cbz group, Fig. 9.7. Initial hydrogenolysis of the O–Bn bond followed by decarboxylation of the carbamic acid so formed.

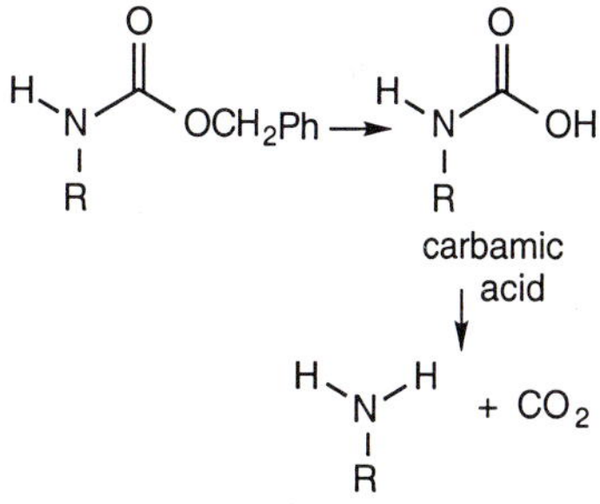

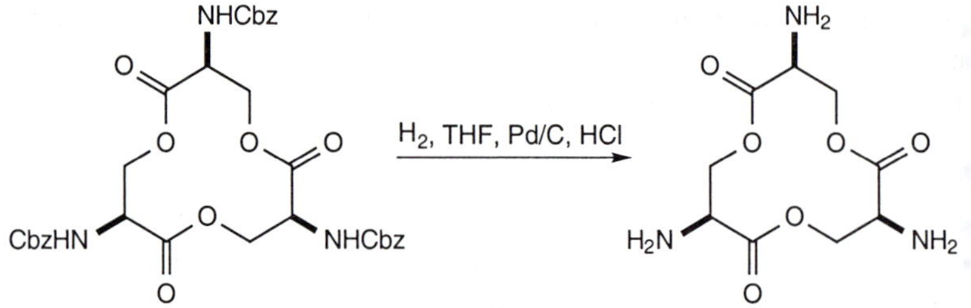

Fig. 9.7

Cleavage of non-benzylic carbon–oxygen bonds

So, it is relatively easy to cleave a benzylic carbon–oxygen bond. Wh about C–O bonds that are not part of a benzylic system? We shall cover th topic by examining ways of deoxygenating various types of alcohols: th process illustrates the most important methods of C–O bond cleavage and frequently used in synthesis.

Deoxygenation of alcohols under ionic conditions

Perhaps the easiest method of deoxygenating alcohols involves formation of good leaving group, such as a toluenesulfonate (OTs) derivative, a subsequent displacement with hydride using $LiAlH_4$. The reaction involv nucleophilic attack of AlH_4^- on carbon (the Li^+ cation acts as a Lewis acid assist the toluenesulfonate group in leaving), Fig. 9.8. As befits an S reaction, this process is sensitive to steric factors and works best wi primary alcohols: secondary alcohols react slowly and tertiary alcohols a clearly inert to displacement and show competing elimination.

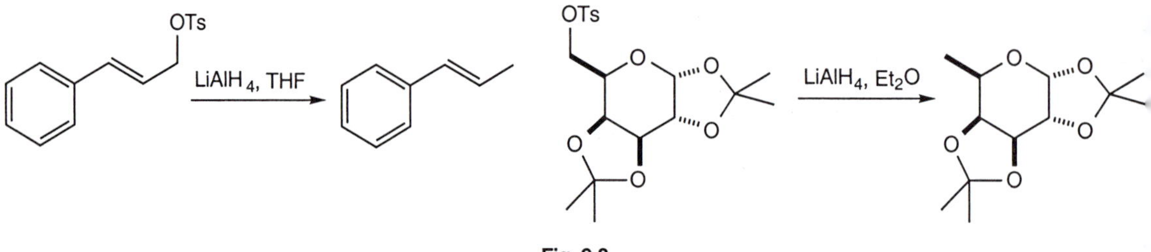

Fig. 9.8

One obvious way of deoxygenating tertiary alcohols would invol elimination to an alkene, followed by hydrogenation. However, a neat on step procedure has been developed that uses hydride transfer from a sila (Et_3SiH is often used) in the presence of acid, Fig. 9.9.

Remember that the silicon–hydrogen bond is polarised $Si^{\delta+}$–$H^{\delta-}$, so delivery of hydride should not come as a surprise. The key addition step is shown below.

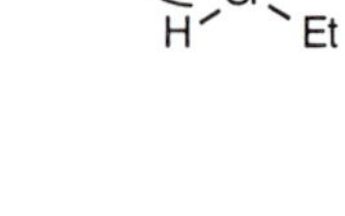

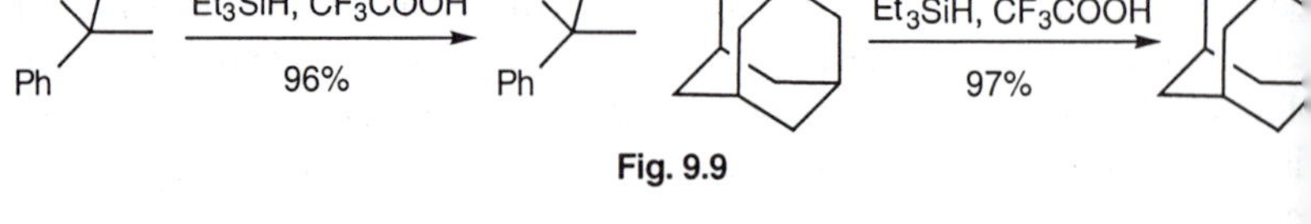

Fig. 9.9

Silanes are not as reactive as boranes or alanes and require a pote electrophile before they will act as a reducing agent. The role of the acid ıis reaction is to protonate the alcohol and promote formation of a tertiary arbocation, Fig. 9.9. Some secondary alcohols will also react under these onditions, so long as they can form stable carbocations.

eoxygenation of alcohols under radical conditions

erhaps the most general method of reductive cleavage of a hydroxyl group ıvolves radicals and is known as the Barton–McCombie reaction. First, lcohols are converted into their xanthate ester derivatives (NaH, CS_2, MeI) ıd these are then reacted with tributyltin hydride in hot solvent such as enzene: the resulting radical chain reaction is shown in Fig. 9.10. Initiators an be added to form small amounts of $Bu_3Sn\bullet$, although the reaction usually roceeds without them.

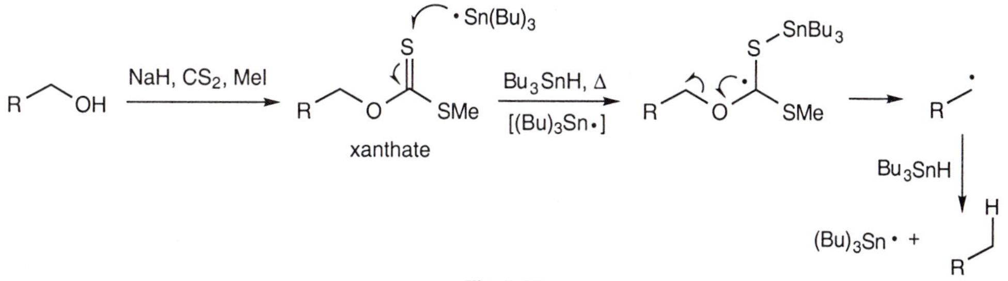

Fig. 9.10

While such radical-mediated deoxygenation reactions work very well for oth primary and secondary alcohols, tertiary alcohols are not normally good ıbstrates because the xanthate ester undergoes a concerted elimination (the hugaev reaction) at the elevated temperatures required for deoxygenation ith tributyltin hydride, Fig. 9.11.

The Chugaev reaction

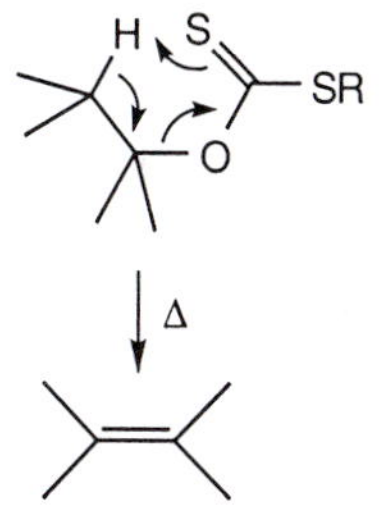

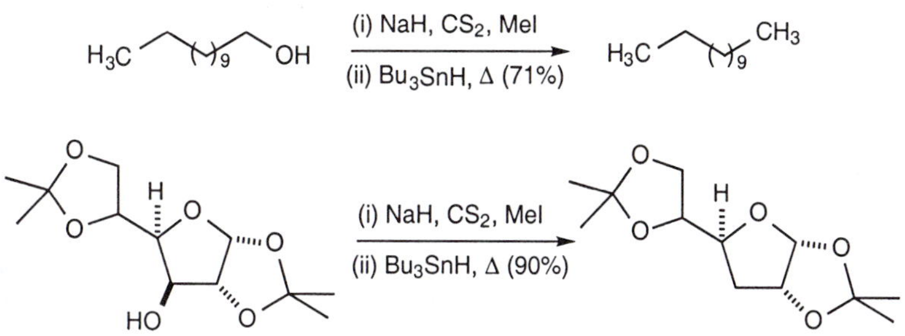

Fig. 9.11

An ingenious modification of the Barton–McCombie reaction allows cinal diols to be reduced *via* formation of a thionocarbonate, Fig. 9.12. eduction of this carbonate with tributyltin hydride is rather similar to that ıown earlier. In this case, after radical addition, the thionocarbonate ıllapses to expel the more substituted (and more stable) carbon radical and so is chemistry is a predictable method of reducing one hydroxyl group from a cinal diol.

Alternatives such as the thionoformates ($R_3COCH{=}S$) may be used to derivatise and deoxygenate tertiary alcohols without competing elimination.

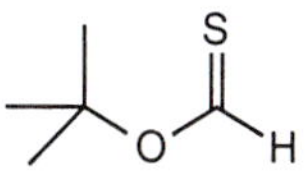

thionoformate

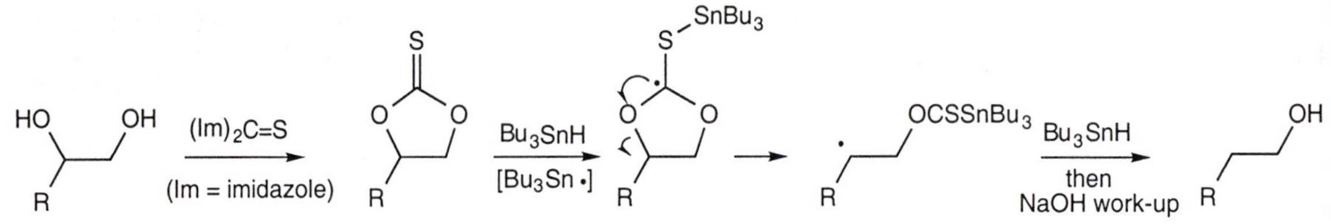

Fig. 9.12

Deoxygenation of epoxides to alkenes

Alkenes may be produced from epoxides *via* a complete deoxygenatio process; clearly this is opposite to the epoxidation of alkenes discussed i Chapter 3. Bearing in mind their strong affinity for oxygen, it should com as little surprise to note that phosphorus reagents are particularly effective fo deoxygenating epoxides; for example, both Ph_3P and $(EtO)_3P$ wi deoxygenate epoxides. A nice method of transforming epoxides to alken with high levels of stereospecificity was reported in the 1970s by E. Vede who used $LiPPh_2$ to open epoxides. Subsequent reaction *in situ* with meth iodide (this reacts with a lone pair on phosphorus rather than the alkoxid which is stabilised by an O–Li bond) forms a betaine. Cyclisation ensues ar extrusion of $MePh_2P{=}O$ forms an alkene (compare with the Wittig reaction).

The reaction is stereospecific because the epoxide opens with stri inversion and the elimination is *syn* periplanar, see the conversion of a *tran* epoxide into a *cis*-alkene, Fig. 9.13 (see also Fig. 2.5, page 8). Limitatior of the $LiPPh_2$ reagent are that it will react as both a nucleophile and a bas with other functional groups such as esters and ketones. Sterically hindere epoxides are also slow to react under these deoxygenation conditions.

Because both the epoxidation of alkenes and the deoxygenation of epoxides are sterospecific, one can use the oxidation/reduction protocol to invert the geometry of an alkene. So, imagine converting a *cis*-alkene to *cis*-epoxide (*m*–CPBA) and thence to a *trans*-alkene with Ph_2PLi/ MeI.

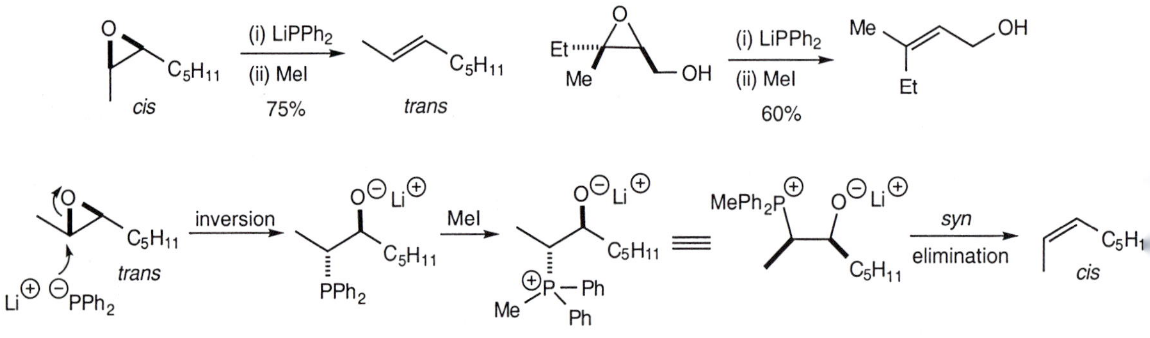

Fig. 9.13

Reductive cleavage of epoxides to alcohols

Epoxides can also be reduced another way, to give alcohols. Reaction of and **5** with $LiAlH_4$ illustrates the addition of AlH_4^- to an epoxide electrophil Fig. 9.14. Generally (as befits an S_N2 type reaction), the reducing age attacks the least substituted end of the epoxide and, as we would expect, th reaction goes with inversion.

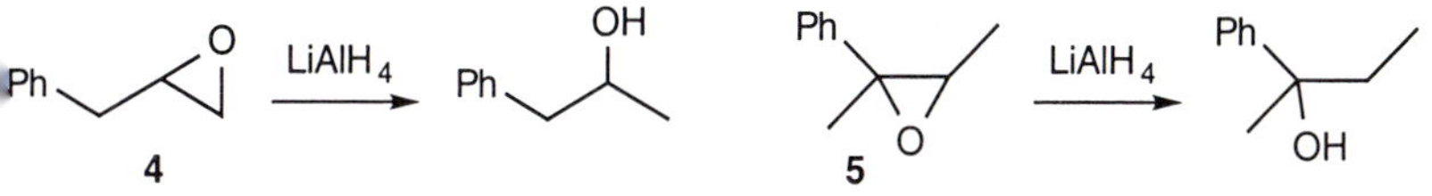

Fig. 9.14

Remember that S_N2 reactions are very sensitive to steric effects, so the less hindered end of the epoxide is attacked first. The lithium cation provides some activation to ring opening.

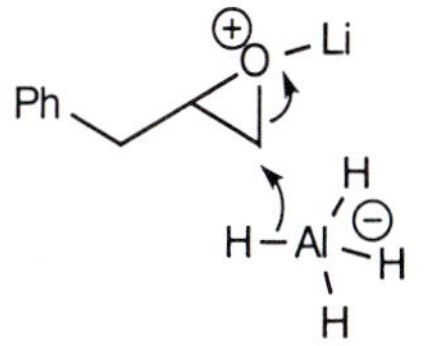

The reduction of epoxy-alcohols by $LiAlH_4$ is an interesting reaction that has proven useful in synthesis. Early studies showed that epoxide **6** was opened regioselectively to give diol **7**, Fig. 9.15. The origins of such regiocontrol lie in the initial reaction of $LiAlH_4$ with the alcohol to form aluminate complex **8**; this is then ideally arranged for nucleophilic attack at the proximal end of the epoxide. Quenching the reaction with H_3O^+ will cleave the O–Li and O–Al bonds to generate diol **7**.

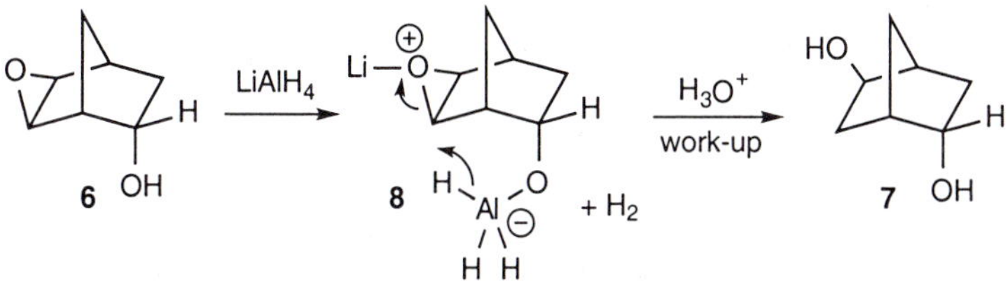

Fig. 9.15

If you have read Chapter 3 then the Sharpless epoxidation reaction will be familiar: it transforms allylic alcohols into epoxy alcohols with high enantiomeric purity. The epoxy alcohols so generated may be reduced rather selectively to give a range of either 1,2- or 1,3-diols depending on the reducing agent chosen and the geometry of the epoxide, Fig. 9.16.

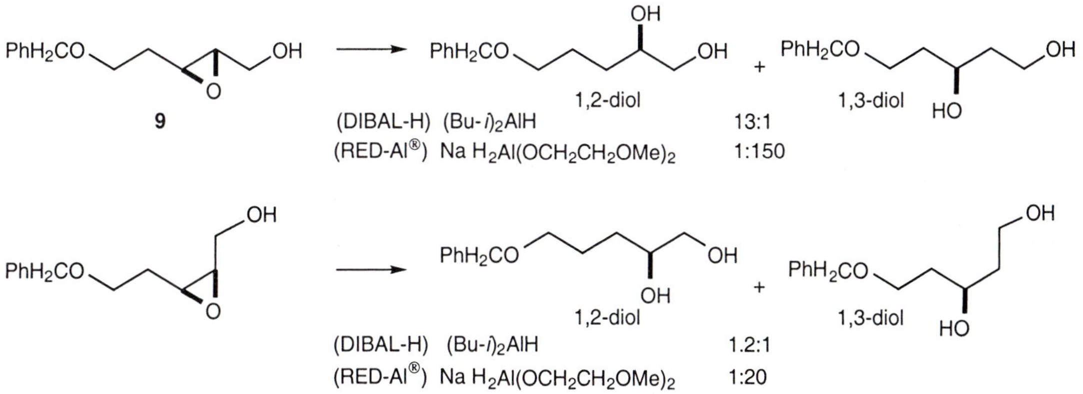

Reducing agent	1,2-diol : 1,3-diol (epoxide **9**)
(DIBAL-H) $(Bu\text{-}i)_2AlH$	13:1
(RED-Al®) $Na\ H_2Al(OCH_2CH_2OMe)_2$	1:150

Reducing agent	1,2-diol : 1,3-diol
(DIBAL-H) $(Bu\text{-}i)_2AlH$	1.2:1
(RED-Al®) $Na\ H_2Al(OCH_2CH_2OMe)_2$	1:20

Fig. 9.16

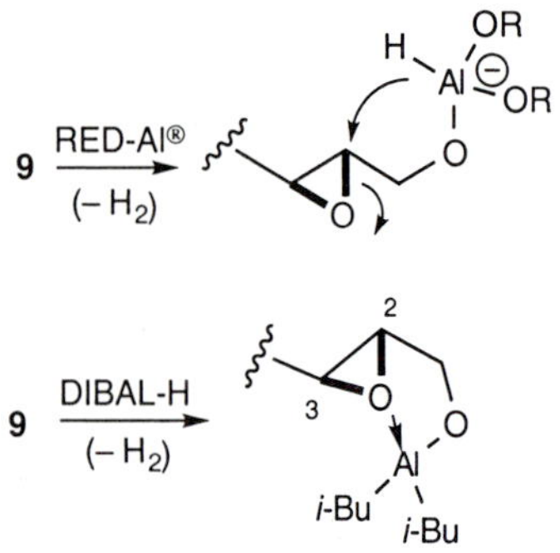

The difference in selectivity with DIBAL-H versus RED-Al® has its origins in the different complexes that each makes with the alcohol, prior to epoxide opening. For example, RED-Al® contains a negatively charged aluminium, like $LiAlH_4$, and will make an ATE complex with an alcohol. This can then deliver a hydride intramolecularly to the nearest carbon atom to produce a 1,3-diol. On the other hand, DIBAL-H will form a neutral complex that cannot deliver hydride in an intramolecular fashion. Subsequent coordination of the Lewis acidic aluminium to the epoxide oxygen seems to promote regioselective intermolecular nucleophilic attack (by more DIBAL-H) at C–3 forming a 1,2-diol. The factors that give rise to this sense of regioselectivity may be stereoelectronic in origin.

Reductive cleavage of a carbon–oxygen double bond

The Wolff-Kishner reduction is an interesting method of reducing aldehydes and ketones to methyl or methylene groups respectively. The original procedure involved reaction of a carbonyl group with hydrazine to form a hydrazone and this was then reacted with hydroxide at elevated temperatures to form the reduction product. Subsequent modifications to this protocol have enabled formation of the hydrazone *in situ* rather than in a separate step. So, the Huang-Minlon modification of the Wolff-Kishner involves reaction of a carbonyl compound, hydrazine, KOH and ethylene glycol at high temperatures, Fig. 9.17.

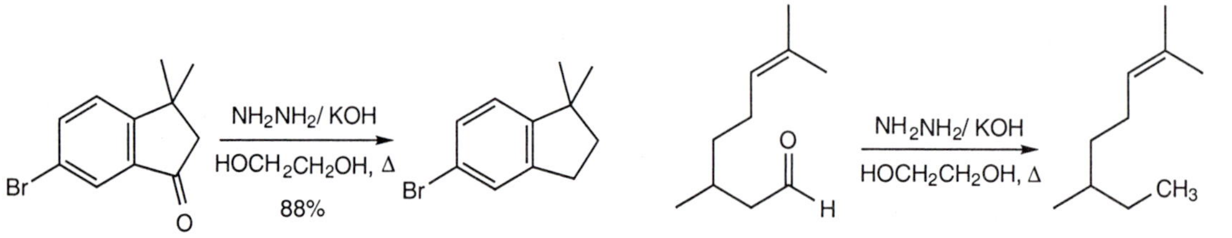

Fig. 9.17

The mechanism of reduction is thought to involve initial deprotonation of the hydrazone, followed by protonation of the anion on carbon, Fig. 9.18. Further deprotonation ensues and the resulting azine will collapse rapidly (and irreversibly) to evolve nitrogen gas and form a carbanion which will be protonated by the glycol.

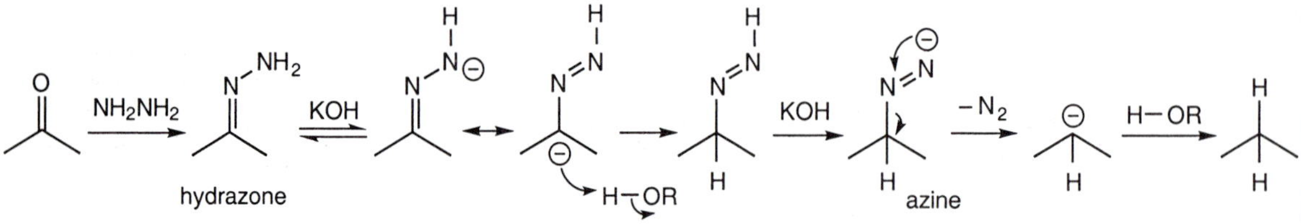

Fig. 9.18

There are other ways to reduce a carbonyl group completely. For example, the Clemmensen reduction employs Zn/Hg in HCl to achieve the same transformation as the Wolff-Kishner reaction. Later on in this chapter you will come across a desulfurization reaction that can be used to reduce carbonyl groups in a two-step process.

9.3 Reductive dehalogenation

Hydrogenolysis of the C–X bond

Halogens can normally be reductively cleaved by hydrogen and a metal catalyst, i.e. hydrogenolysis. Palladium is perhaps the best metal for this purpose and a base is normally added to the reaction as the HX by-product from cleavage may retard the rate of reduction, Fig. 9.19.

Generally, the ease of hydrogenolysis follows the same order as carbon–halogen bond strengths, C–I < C–Br < C–Cl < C–F, with the weakest bond, C–I, being easiest to reduce.

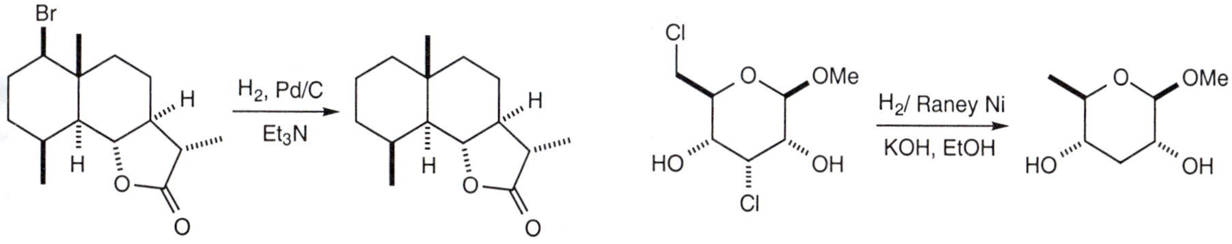

Fig. 9.19

All sorts of halides may be reduced, including those bound to trigonal carbon, Fig. 9.20. Of course, hydrogenolysis of a vinyl halide can also lead to reduction of the alkene unit: in some cases successful hydrogenolysis without hydrogenation can be accomplished with Lindlar's catalyst.

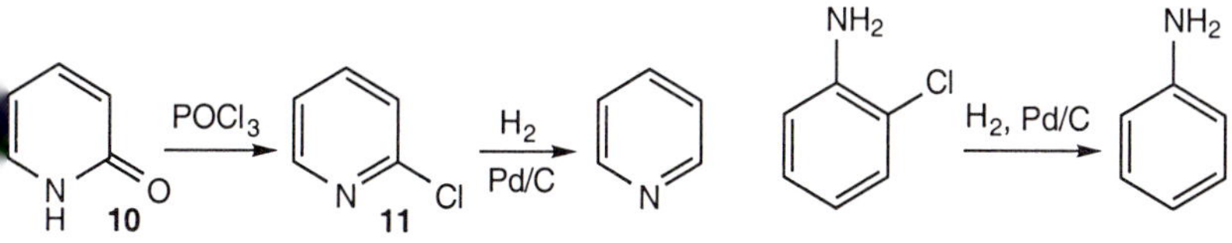

The conversion of **10** into **11** followed by hydrogenolysis is a good way of reducing pyridones.

Fig. 9.20

Hydrogenolysis of acid chlorides (themselves readily prepared from carboxylic acids and $(COCl)_2$) yields aldehydes and is known as the Rosenmund reduction, Fig. 9.21. Again, palladium is the metal of choice and this catalyst usually needs deactivating prior to reduction. In this case, simply heating the metal in xylene under an atmosphere of H_2 achieves sufficient deactivation, probably by changing the surface of the catalyst.

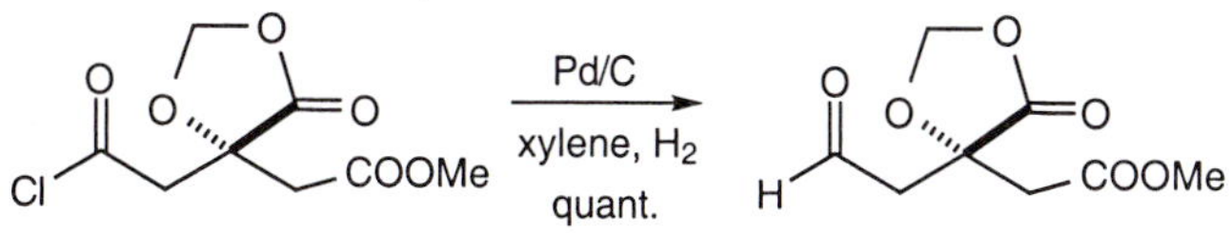

Problems sometimes associated with the Rosenmund reduction are hydrolysis of the acid chloride, over-reduction of the aldehyde to an alcohol and decarbonylation.

Fig. 9.21

Dehalogenation by radical-based reagents

Halogens may also be removed from organic compounds *via* radical reactions. Tributyltin hydride is the best reagent and this reduction is related to the

Barton–McCombie reaction described earlier. The conditions involve reaction of an organo-halide with ≥1 eq. of Bu_3SnH and a small amount of radical initiator (AIBN: see Fig. 4.14, page 30 for the structure of AIBN and the radical it forms upon thermolysis) in an inert solvent such as benzene, Fig. 9.22.

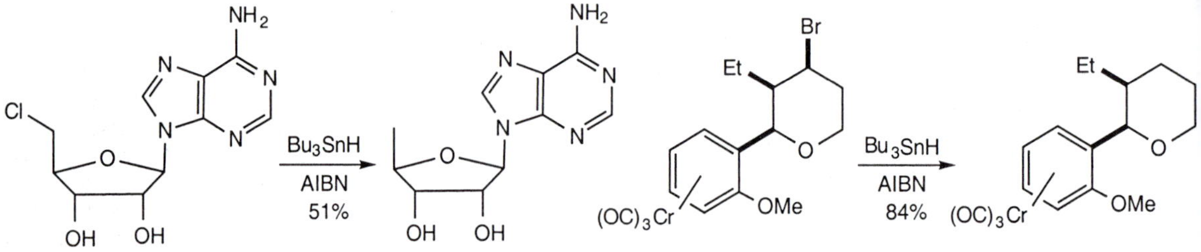

Fig. 9.22

Tin residues can be difficult to remove completely from these types of reaction and a variant of this reaction has appeared that utilises catalytic Bu_3SnCl and $NaBH_4$ to continually generate Bu_3SnH *in situ*, Fig. 9.23.

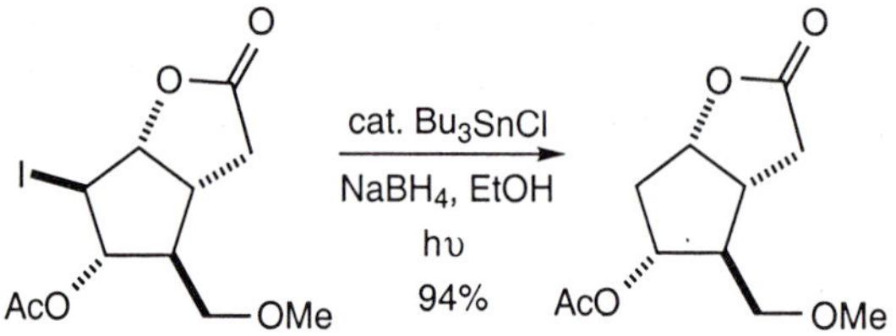

Fig. 9.23

The order of reactivity for Bu_3SnH mediated reduction is RI > RBr > RCl > RF. Tertiary halides are also more reactive than secondary which in turn are more reactive than primary halides. Aryl and vinyl halides are even less reactive again; for example iodobenzene is only reduced to benzene in neat Bu_3SnH at 120°C.

The mechanism of the Bu_3SnH mediated reduction requires initial formation of a radical from the initiator (denoted as X• in Fig. 9.24); this can then react with tin hydride to produce a tributyltin radical. Halogen atom abstraction ensues (the driving force is formation of the strong tin–halogen bond) and the carbon-based radical that is formed as a result can propagate the chain by abstracting a hydrogen atom from more tributyltin hydride.

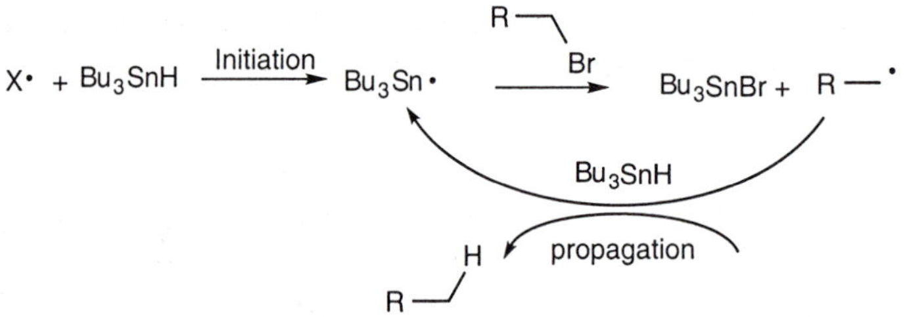

Fig. 9.24

Bearing in mind the order of reactivity mentioned earlier, it is possible to discriminate between two dissimilar halogens within the same compound, Fig. 9.25 (the weaker carbon–halogen bond is cleaved). Of course, it is also possible to selectively reduce halogens that give rise to stabilised radicals, see **12→13**.

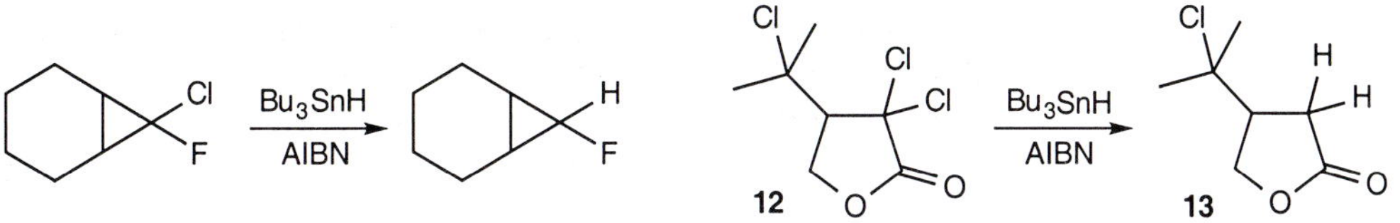

Fig. 9.25

Reductive fission of alkyl halides using reactive metals

Some metals are effective at reducing carbon–halogen bonds *via* electron-transfer processes. Reductive cleavage involves addition of an electron to a C–X bond followed by cleavage to form a carbon-based radical and a halide anion, Fig. 9.26. Further reduction of the radical to a carbanion ensues and this may then be protonated *in situ*.

Of course, one could draw the bond cleavage (Fig. 9.26) to give a carbanion and X• which would then be reduced to X^-. In this case it seems sensible to form the anion on the more electronegative atom (i.e. X). Although in some systems the particular mode of bond cleavage is debatable, you should remember that after addition of two electrons, the net outcome is the same.

$$R_3C-X \xrightarrow{+e^-} R_3C\cdot + X^{\ominus} \xrightarrow{+e^-} R_3C^{\ominus} + X^{\ominus} \xrightarrow{H^+} R_3C-H$$

Fig. 9.26

Several metals are effective in this role and all are capable of being electron donors (e.g. Li, Na, Mg, Zn, Sm). So, reduction of the following organohalogen compounds can be achieved under a variety of different conditions, Fig. 9.27, which all consist, essentially, of a source of electrons (the metal) combined with a source of protons (e.g. ROH, NH_3, HOAc).

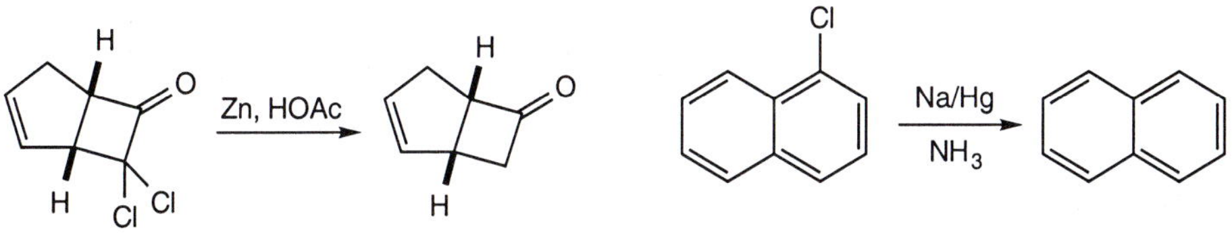

Fig. 9.27

The reductive cleavage of a halogen adjacent to a carbonyl group is particularly favourable. In these cases, one may draw a mechanism which involves direct addition of electrons to the C=O, followed by expulsion of a leaving group to form a stabilised radical. This radical can then be reduced by addition of another electron and protonation of the resulting enolate completes the sequence and forms the reduced product, Fig. 9.28.

This particular reduction works well for the cleavage of a host of leaving groups other than the halogens (i.e. X= OH, OAc, PhS).

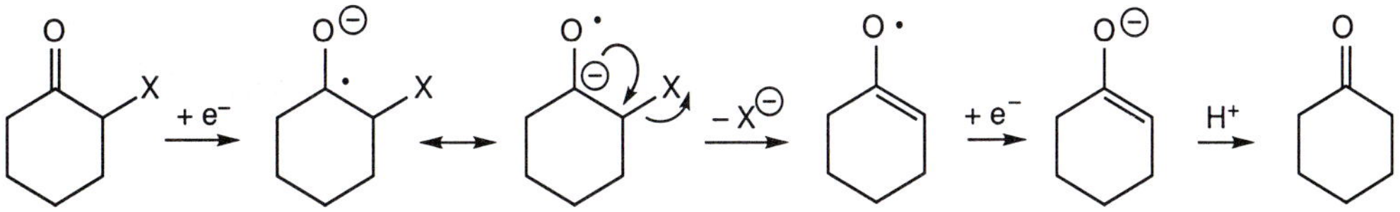

Fig. 9.28

9.4 Reductive cleavage of carbon–sulfur bonds

Consider the reductive cleavage of carbon–halogen bonds shown in Fig. 9.27; then it should come as little surprise to note that carbon–sulfur bonds in sulfides, sulfoxides and sulfones may also be cleaved with a reactive metal and a suitable proton source, Fig. 9.29.

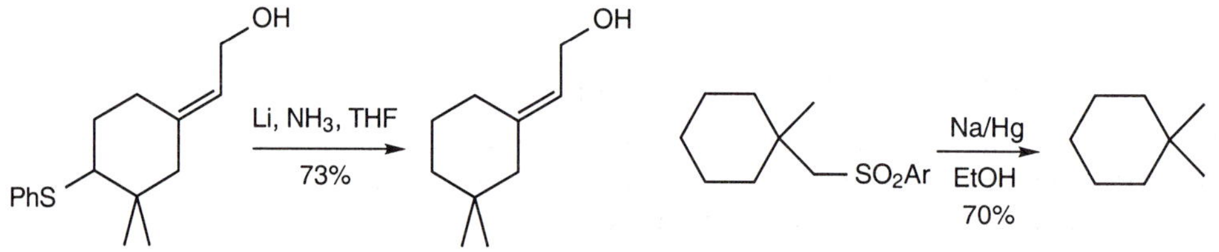

Li, NH_3, THF 73%

Na/Hg EtOH 70%

Fig. 9.29

The preparation of Raney nickel leaves the metal with hydrogen already adsorbed onto the surface. Therefore, reactions using this catalyst need not be performed under an atmosphere of hydrogen.

The reductive cleavage of carbon–selenium bonds can also be accomplished with Raney nickel.

One of the best ways to completely reduce a sulfur-containing group is to use Raney nickel and Fig. 9.30 shows the hydrogenolysis of sulfides, sulfoxides and sulfones. Raney nickel consists of fine particles of nickel metal that are prepared from a nickel/aluminium alloy that is heated with sodium hydroxide. Raney nickel of differing activity may be produced by variation in the conditions used during preparation. Although active Raney nickel will reduce carbon–carbon multiple bonds, one can often cleave carbon–heteroatom bonds without hydrogenation by using a deactivated form of the catalyst.

Conversion of a carbonyl group to a thioacetal, followed by hydrogenolysis with Raney nickel, is another way to completely reduce a C=O under essentially neutral conditions (compare with the basic conditions of the Wolff-Kishner and the acidic conditions of the Clemmensen reduction).

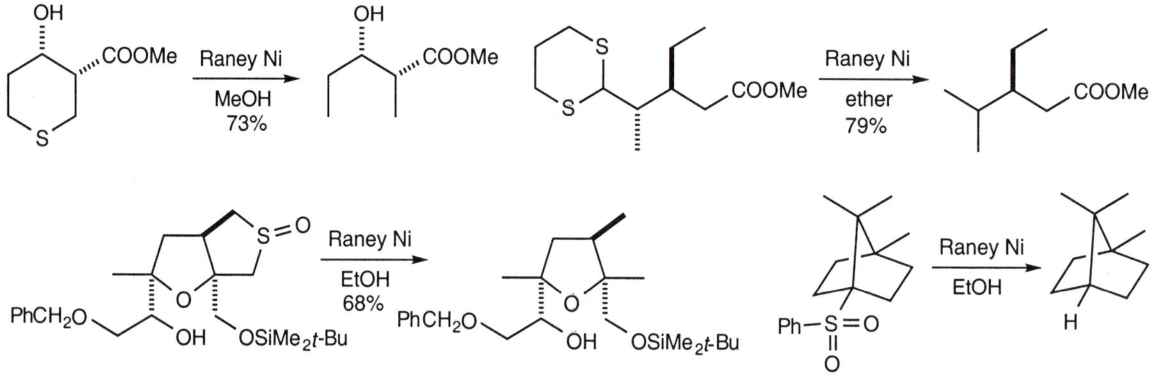

Raney Ni MeOH 73%

Raney Ni ether 79%

Raney Ni EtOH 68%

Raney Ni EtOH

Fig. 9.30

Finally, it is worth mentioning that radical reactions using Bu_3SnH may be used to reduce sulfides to hydrocarbons, Fig. 9.31.

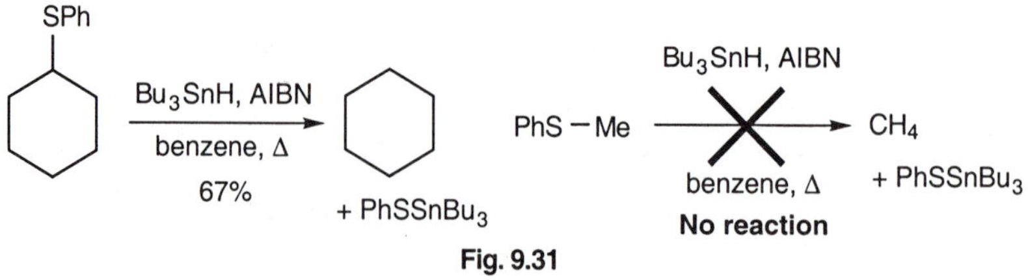

Bu_3SnH, AIBN benzene, Δ 67% + $PhSSnBu_3$

PhS – Me Bu_3SnH, AIBN benzene, Δ **No reaction** → CH_4 + $PhSSnBu_3$

Fig. 9.31

This reaction works best if the radical formed from C–S bond cleavage is tertiary or stabilised by conjugation. The mechanism for the reduction is analogous to that shown earlier for reduction of carbon–halogen bonds.

9.5 Reductive cleavage of cyclopropane rings

The partial π-character of C–C bonds within cyclopropane rings means that they may be cleaved by hydrogenolysis, Fig. 9.32. With alkyl substituents present on the ring, it tends to be the least hindered bond that is broken; this regioselectivity can be used to prepare geminal dimethyl groups.

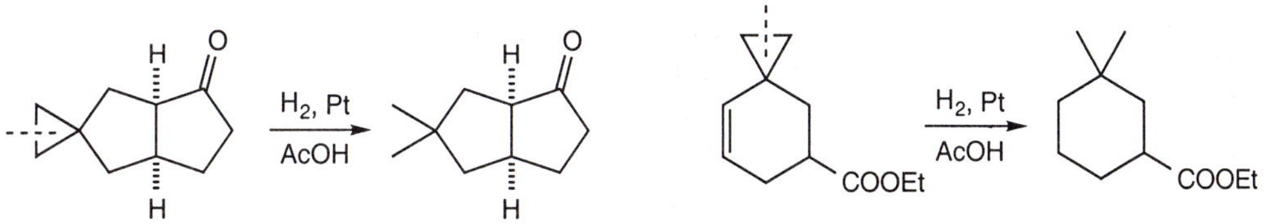

Fig. 9.32

It has been postulated that rupture of the cyclopropane ring takes place on the catalyst surface and a doubly adsorbed species is formed. Presumably, transfer of hydrogen from the metal to the carbons ensues to form the geminal dimethyl group.

Cyclopropanes that are activated by virtue of being adjacent to a carbonyl group can also be cleaved with reactive metals such as sodium or lithium in ammonia, Fig. 9.33.

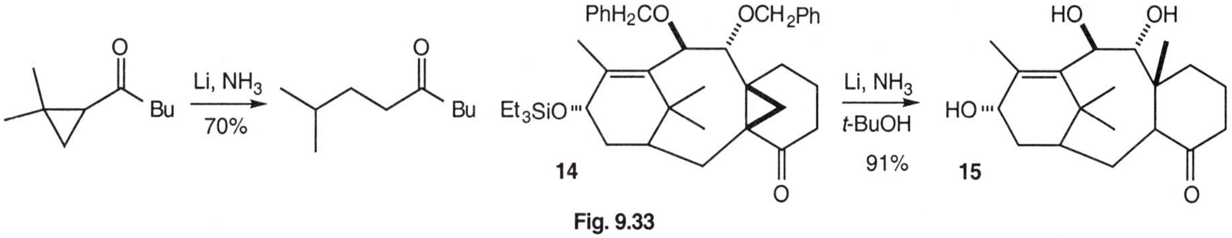

Fig. 9.33

In some cases, the most substituted cyclopropane bond is cleaved during ring opening, forming the most stable radical. However, other factors may influence the regioselectivity of this ring cleavage, most notably the ease of overlap between the π-system of the radical anion and the relevant orbitals of the cyclopropane, see **14**→**15**.

The mechanism probably starts with addition of an electron to the electron-deficient carbonyl group to form a radical anion; this may then cleave to expel a carbon-based radical and form an enolate, Fig. 9.34. Under these reducing conditions, the carbon-based radical will be reduced further to an anion and, finally, protonation of both the anion and the enolate (either *in situ* or with an external proton quench) forms the observed product.

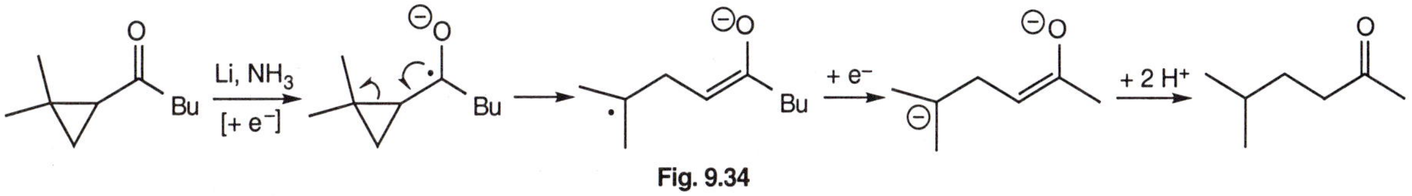

Fig. 9.34

Further reading

Oxidation

Haines, A. H. (1985). *Methods for the oxidation of organic compounds: alkanes, alkenes, alkynes, and arenes*, Academic Press, London.

Haines, A. H. (1988). *Methods for the oxidation of organic compounds: alcohols, alcohol derivatives, alkyl halides, nitroalkanes, allyl azides, carbonyl compounds, hydroxyarenes and aminoarenes*, Academic Press, London.

Trost, B. M. and Fleming, I. (ed.) (1991). *Comprehensive organic synthesis*, Pergamon, Oxford. Vol. 7.

Reduction

Freifelder, M. (1978). *Catalytic hydrogenation in organic synthesis procedures and commentary*, Wiley, New York.

Hudlicky, M. (1984). *Reductions in organic chemistry*, Ellis Horwood, Chichester.

Rylander, P. N. (1985). *Hydrogenation methods*, Academic Press, London.

Bartók, M. (1985). *Stereochemistry of heterogeneous metal catalysis*, Wiley, Chichester.

Trost, B. M. and Fleming, I. (ed.) (1991). *Comprehensive organic synthesis*, Pergamon, Oxford. Vol. 8.

Seyden–Penne, J. (1997). *Reductions by the alumino– and borohydrides in organic synthesis*, Wiley–VCH, New York.

General guides to oxidation and reduction

Excellent chapters on the redox properties of organic compounds may be found in the following.

House, H. O. (1972). *Modern synthetic reactions*, W. A. Benjamin, California.

March, J. (1985). *Advanced organic chemistry*, Wiley, New York.

Carruthers, W. (1986). *Some modern methods of organic synthesis*, Cambridge University Press, Cambridge.

Smith, M. B. (1994). *Organic synthesis*, McGraw-Hill, New York.

Burke, S. D. and Danheiser, R. L. (ed.) (1999). *Handbook of reagents for organic sythesis: oxidising and reducing agents*, Wiley, Chichester.

Index